The Dynamic Cosmos

Exploring the physical evolution of the universe

Mark S. Madsen

CHAPMAN & HALL
London · Glasgow · Weinheim · New York · Tokyo · Melbourne · Madras

Published by Chapman & Hall, 2–6 Boundary Row, London SE1 8HN, UK

Chapman & Hall, 2–6 Boundary Row, London SE1 8HN, UK

Blackie Academic & Professional, Wester Cleddens Road, Bishopbriggs, Glasgow G64 2NZ, UK

Chapman & Hall GmbH, Pappelallee 3, 69469 Weinheim, Germany

Chapman & Hall USA, 115 Fifth Avenue, New York, NY 10003, USA

Chapman & Hall Japan, ITP-Japan, Kyowa Building, 3F, 2-2-1 Hirakawacho, Chiyoda-ku, Tokyo 102, Japan

Chapman & Hall Australia, 102 Dodds Street, South Melbourne, Victoria 3205, Australia

Chapman & Hall India, R. Seshadri, 32 Second Main Road, CIT East, Madras 600 035, India

First edition 1995

© 1995 Mark S. Madsen

Printed in Great Britain by TJ Press Ltd, Padstow, Cornwall

ISBN 0 412 62300 5

Apart from any fair dealing for the purposes of research or private study, or criticism or review, as permitted under the UK Copyright Designs and Patents Act, 1988, this publication may not be reproduced, stored, or transmitted, in any form or by any means, without the prior permission in writing of the publishers, or in the case of reprographic reproduction only in accordance with the terms of the licences issued by the Copyright Licensing Agency in the UK, or in accordance with the terms of licences issued by the appropriate Reproduction Rights Organization outside the UK. Enquiries concerning reproduction outside the terms stated here should be sent to the publishers at the London address printed on this page.
The publisher makes no representation, express or implied, with regard to the accuracy of the information contained in this book and cannot accept any legal responsibility or liability for any errors or omissions that may be made.

A catalogue record for this book is available from the British Library

Library of Congress Catalog Card Number: 95-70841

∞ Printed on acid-free text paper, manufactured in accordance with ANSI/NISO Z39.48-1992 (Permanence of Paper).

To Deborah Lea Madsen

CHAPMAN & HALL MATHEMATICS SERIES

EDITORS:

Professor Keith Devlin
St Mary's College
Moraga, California
USA

Professor Derek Goldrei
Open University
Oxford
UK

Dr James Montaldi
Université de Nice
Nice
France

OTHER TITLES IN THE SERIES INCLUDE

Functions of Two Variables
S. Dineen

Network Optimization
V. K. Balakrishnan

Sets, Functions and Logic
A foundation course in mathematics
Second edition
K. Devlin

Algebraic Numbers and Algebraic Functions
P. M. Cohn

Dynamical Systems
Differential equations, maps and chaotic behaviour
D. K. Arrowsmith and C. M. Place

Control and Optimization
B.D. Craven

Elements of Linear Algebra
P. M. Cohn

Full information on the complete range of Chapman & Hall mathematics books is available from the publishers.

Contents

Foreword	ix
Preface	xi
Acknowledgements	xv

1 Introduction — 1
  1.1 The Purpose of Cosmology — 1
  1.2 Structure of This Book — 1

2 The Observed Universe — 5
  2.1 Introduction — 5
  2.2 Existence of Structure — 6
  2.3 Cosmic Expansion — 6
  2.4 The Age of the Cosmos — 8
  2.5 Element Abundances — 10
  2.6 Cosmic Matter — 10
  2.7 Background Radiation — 12
  2.8 Mass Density — 13
  2.9 Isotropy and Homogeneity — 14
  2.10 Evidence for Evolution — 15
  2.11 Causal Horizons — 16
  2.12 Review — 17
  2.13 References — 18

3 The Cosmological Equations — 21
  3.1 The Dynamical Equations from Cosmic Energy — 21
  3.2 Derivation and Explanation of Hubble's Law — 25
  3.3 Solutions to the Expansion Equations — 26

3.4	Review	30
3.5	Exercises	30
3.6	References	31

4 Cosmological Redshift and Horizons   33
    4.1 Theory of the Redshift   33
    4.2 Cosmological Horizons   36
    4.3 Review   38
    4.4 Exercises   38
    4.5 References   38

5 Evolution of the Cosmological Density Parameter   39
    5.1 The Flatness of the Universe   39
    5.2 The Flatness Approximation   41
    5.3 Relation of Density and Horizons   42
    5.4 The Evolution of the Density Parameter   43
    5.5 Review   45
    5.6 Exercises   46
    5.7 References   47

6 The Thermal History of the Universe   49
    6.1 The State of Matter in the Early Universe   49
    6.2 The Cosmic Entropy Density   50
    6.3 How the Universe Cooled   52
    6.4 The Neutrino Temperature   54
    6.5 Decoupling and Recombination   56
    6.6 Review   57
    6.7 Exercises   58
    6.8 References   58

7 Cosmological Synthesis of Elements   59
    7.1 Time of Element Synthesis   59
    7.2 The Neutron to Proton Ratio   60
    7.3 The Helium Mass Fraction   62
    7.4 Review   64
    7.5 Exercises   65
    7.6 References   65

8 Cosmic Asymmetry and the Origin of Matter   67
    8.1 The Mystery of the Existence of Matter   67
    8.2 Asymmetry as a Physical Theory   68
    8.3 The Generation of Baryon Number   69
    8.4 Review   70
    8.5 Exercises   70

8.6	References	70

9 Primordial Cosmological Inflation — 71
- 9.1 The Cosmological Problems — 71
- 9.2 Matter Fields in the Early Universe — 72
- 9.3 The Chaotic Inflationary Model — 75
- 9.4 The Planck Limit — 77
- 9.5 Density Fluctuations Generated by Inflation — 78
- 9.6 The End of Inflation — 80
- 9.7 Dissipation of Primordial Inhomogeneity — 80
- 9.8 Eternal Chaotic Inflation — 81
- 9.9 Review — 83
- 9.10 Exercises — 83
- 9.11 References — 84

10 The Evolution of Cosmic Structure — 85
- 10.1 Matter as a Fluid — 85
- 10.2 The Evolution of Small Perturbations — 86
- 10.3 Gravitational Energy and the Jeans Scale — 89
- 10.4 Fluctuations of the Background Radiation — 93
- 10.5 Review — 95
- 10.6 Exercises — 96
- 10.7 References — 96

11 Dark Matter and Structure Formation — 97
- 11.1 Evidence for Dark Matter — 97
- 11.2 The Silk Scale — 99
- 11.3 Hot Dark Matter — 101
- 11.4 Cold Dark Matter — 104
- 11.5 Review — 107
- 11.6 Exercises — 108
- 11.7 References — 108

12 Exotic Objects — 109
- 12.1 Black Holes — 109
- 12.2 Magnetic Monopoles and the Monopole Problem — 112
- 12.3 Cosmological Formation of Boson Stars — 116
- 12.4 Review — 119
- 12.5 Exercises — 121
- 12.6 References — 121

13 Survey of Cosmological Theory — 123
- 13.1 What Have We Learned? — 123
- 13.2 The Present State of Cosmological Knowledge — 125

13.3 Problems Currently Facing Cosmology	125
13.4 Reprise	126

Appendix A Physical Units and Constants — 127

Appendix B Selected Bibliography of Cosmology — 131

B.1 Introductory References	131
B.2 Intermediate References	132
B.3 Advanced References	133
B.4 Relativistic Cosmology	134
B.5 Selected Journal Articles	135
B.6 General Physics	136
B.7 Complete List of Articles	137
B.8 Complete List of Books	139

Index — 140

Foreword

Cosmology is one of the most spectacular successes of the human mind. By the careful study of the local manifestations of gravity, Newton and Einstein have bequeathed us a set of mathematical rules which describe all observed effects of gravity to extraordinary precision. Moreover, Einstein's theory of gravitation provides us with the means to describe the changing structure of entire universes. Remarkably, one of the simplest candidates in this cosmic gallery appears to be that universe in which we find ourselves. Modern cosmology is the quest to pin down the structure of this universe with ever-greater precision, and to uncover its past history as far back in time as time will go. But that is not all that cosmology is. Cosmologists seek understanding as well as knowledge. Where the universe displays unusual structure, we seek an explanation for it: why are there things like galaxies, why do they have the sizes and clustering patterns that we see? Why does the universe expand at a similar rate in every direction? Why is there so little antimatter between the stars? Why does most of the material in the universe reside in forms that are dark and unseen by optical telescopes? And, what is the identity of this mysterious dark matter? These are some of the questions that modern cosmologists seek to answer. In their quest they have found that the study of the largest structures in the universe requires an understanding of the most elementary particles of matter, entities whose properties controlled events during the first fraction of a second of the universe's existence. It was during these opening moments of the universe's history that the conditions were set for the slowly unfolding panoply of galaxies, stars, and planets that would follow billions of years later. Remarkably, many of these new developments in our picture of the universe, and the observational evidence that underpins them, are within the reach of undergraduate students of physics. This new book is an attempt to introduce them to the study of modern cosmology in a manner that stresses the application of simple physics and mathematics to the study of the past and present structure of the universe. By avoiding the use of general relativity, and introducing only those properties of elementary particles that are really necessary to understand the environment in which they appear, Mark Madsen enables students of physics to go beyond popular qualitative descriptions of cosmology and see how the simple principles of physics, so painfully learnt elsewhere, open up an important arena of modern science. By charting a course between the popular introductions to cosmological ideas and technical monographs aimed at the research worker, the author has made a valuable contribution to the expansion of knowledge about the expanding universe.

John D. Barrow
Professor of Astronomy
Astronomy Centre
University of Sussex

Preface

Humans must always have struggled to understand the universe and their position in it. Historically, we have received many myths and accounts of creation from our predecessors, and no matter how widely separated either geographically or temporally those people were, the main thrust of their construction of the cosmos was to make sense of the vast complexity of the encompassing creation. And now it is our turn. We are the people of the twentieth century, and our understanding of the cosmos is necessarily shaped by our own history. Science is now the dominant influence on our thought and our lives, even when we cannot see it, or do not realize it, so it is inevitable and natural that our particular cosmos should be constructed scientifically.

In previous eras of humanity, the dominant effects were those of magic, and the universe was normally seen as a magical construction whose evolution was controlled by the great magical powers of water and stars, of spirits and wishes and dreams. In the present era, however, science has accomplished feats surpassing even the greatest claimed for the eras of magic, and so magic has become diminished in comparison with science. The cosmos constructed by the use of scientific reasoning is therefore less miraculous than the old universes were, but it is also more rational, more detailed, and possessed of incomparably greater intricacy of design.

The purpose of this book is to explicate the modern form and interplay of the various elements that make up the rational component of our universe – the parts which can be modelled, understood, and even altered by our own mental processes, using the rational, mathematical and physical tools of modern science.

It has been my constant aim, during the writing of this book, to make it as accessible as possible. Clearly, this aim is in tension with the aim of displaying the most important details of current cosmological theory in their mathematical clothing. I have also refused to fend off the necessity for quantitative physical explanations of the physics presented in this study.

I have chosen to minimize this tension by keeping the requisite mathematics to an easily accessible level. Nearly all of the mathematics involves nothing more complicated than ordinary differential equations, although some partial differential equations and vector analysis are unavoidable if one is to bring out the rich detail of the formation of structure in the cosmos. The main physical relations required are only the simplest and most fundamental laws of universal gravitation (the inverse square force law), thermodynamics (the first law, that of energy conservation), and quantum theory (the uncertainty principle). All the rest of the physics necessary to describe and understand the structure of our universe will be derived and explained in the course of the text. I have also often resorted to the use of dimensional arguments in order to derive useful relations.

A benefit of such an approach is that I have been able to avoid any reliance on general relativity and quantum field theory, but at the same time I have been able to give physical derivations of the evolution equations for the simplest cosmological models. Although these derivations may be criticized as being insufficiently general in the mathematical sense, there are two defences which can be raised. The first is that they are simple, understandable, memorable, and illustrative. The second is that, in many cases, the reason why we believe in the correctness of general relativity is precisely because there are physically simple cases which we can verify agree with the results obtained by more straightforward and intuitively physical methods.

Overall, I believe that I have succeeded in my aim, although possibly to different degrees in different sections. I feel, though, that I should give some advice to the reader who is new to the delights of the study of cosmology, and finds all the equations somewhat off-putting. When reading this book – or any other in the same vein – your interest is more likely to survive the initial encounter with the material if you can remember that every equation is only a shorthand for a set of ideas: these ideas are often described in more detail in the accompanying text, so if you find yourself becoming alienated from the section you are reading, skip over the equations and emphasize the text in your reading. Ask yourself what the model is about, how it is supposed to work, what are the interrelations between its various parts, and why it should be a convincing explanation of the way the universe appears to us. If you persevere in this manner, you will find that the mathematical formulation will come to make sense more rapidly on a second or third reading.

I also feel it necessary to explain that cosmology is a subject far larger and more complex than can possibly be fitted into a book as small and modest as this one. What is more, every aspect of cosmology is still a topic of ongoing research. That means that neither this book nor any other can be taken as the final word on the subject. At the same time, I have tried to select the most robust material on cosmology. That is, I have omitted

Preface

topics which are still undergoing major revisions of opinion and have aimed at selecting those parts of cosmological theory which are most necessary to the coherence of the theoretical structure and which have best stood the tests of analysis and observation.

The universe is a dynamic entity, evolving, developing, and changing in ways that we can hope only partly to comprehend. I hope that this book will demonstrate to the reader that the physical structure and history of the cosmos is beginning to be understood. The future development of our understanding will depend on the quality of the questions which we are able to ask of the wonderful universe in which we live.

Acknowledgements

In the process of writing this book, I have received assistance, encouragement, and support from many sources, including a few surprising ones. This is where I express my thanks to the many who have helped me in some way. All of those mentioned below have earned them many times over.

My friends, collaborators and colleagues in the exciting field of cosmology research, many of whom have helped me to understand the ideas expressed in this book in simpler ways than I could have managed on my own.

The people who read and commented on various drafts of this book: in particular Nigel Berman, Jackie Butcher, Steven Hayles, Robert Low, David Matravers, all of whom took the time to make corrections and improvements. Thanks for your illegible marginal scrawls!

Derek Raine, for discussing the project with me in detail and making more helpful suggestions than I have been able to use.

My editor, Achi Dosanjh, for being unfailingly cheerful and helpful in assisting me with the details of production.

John Barrow, for taking the time from his globetrotting lifestyle to contribute the Foreword.

British Rail, for running the trains from Brighton to the University of Sussex slowly and unreliably enough that I had sufficient spare time on the train to sketch this work in a year of commuting.

My wife, Deborah Madsen, to whom this book is deservedly dedicated, has applied her editorial skills to support this project in numerous concrete ways.

1

Introduction

1.1 The Purpose of Cosmology

Cosmology is the study of the universe – the word derives from the Greek κόσμος (*cosmos*, meaning the universe) and λόγος (*word*, meaning knowledge), so that cosmology is the discipline of seeking knowledge about the universe.

The aim of studying the universe is to satisfy our natural curiosity about the environment in which we find ourselves existing. This curiosity is not new: it is unlikely that humanity is more interested in cosmology now than, say, five millennia ago. If we differ from our forebears, it is in the kinds of questions we think it natural to ask about the universe.

The modern study of cosmology began in the early years of this century, with the formulation of the equations of general relativity by Einstein, and their application to the first cosmological models, initially by De Sitter and later by Friedmann. Since then, the cosmological paradigm that the universe should be explainable in terms of known physical mechanisms and processes has largely held sway.

In this book, the focus is on visualizing and explaining the physical processes which are now believed to have taken place in previous stages of the evolution of the universe. The universe, once believed to be eternal and unchanging, is now interpreted through the observations that we make of it, and the theory that is built on those observations, as being a place of continous change, whose history has contained many different and interesting epochs and events. The material presented in succeeding chapters describes the dynamical and physical history of the cosmos in which we now live.

1.2 Structure of This Book

After this Introduction, the rest of the book is laid out as follows. Chapter 2 surveys and describes the known facts about the universe which must be incorporated into any useful physical theory of the dynamical evolution of the universe.

Chapters 3–5 are concerned with the equations which describe the expansion of the universe. These equations are derived on the basis of energy considerations in Chapter 3, and are then used to study the cosmological redshift and the existence of horizons in Chapter 4. The evolution of the cosmological density parameter is then derived and studied in Chapter 5.

The next few chapters tell the story of successive epochs in the history of the hot universe. Chapter 6 is concerned with studying the thermal history of the hot universe. Chapter 7 then tells the fascinating story of how about one quarter of the nucleon mass was formed into helium nuclei in the first few seconds of the universe's history. Chapter 8 moves further back in time to explain how the nucleons themselves were created by the decay of heavy particles at the earliest stages and highest temperatures. Cosmological inflation – the story of how the universe came to be in the hot state, and how it became so smooth – is the subject matter of Chapter 9, which then goes on to speculate how the universe may have started its expansion in the first place.

The intertwined themes of the formation of large scale structure and the possible existence of dark matter make up the material presented in Chapters 10–11. Chapter 10 examines the behaviour of small perturbations of the density in an expanding universe, while Chapter 11 presents the evidence for the existence of dark matter, and then goes on to describe some of the features present in the structure formation scenarios that result when the dark matter is either hot or cold.

Chapter 12 contains some material that is harder than that presented in the rest of the book. Topics covered there are exotic objects such as superheavy magnetic monopoles, black holes, and boson stars – cold stars consisting of scalar bosonic particles. While not part of the mainstream study of cosmology, these topics are reasonably accessible, have some impact on theories of the evolution of the universe, and illustrate how many of the ideas presented in earlier sections can be applied to the study of new ideas in cosmology.

The text of this book concludes with a survey in Chapter 13, which looks back over the material presented in this book and summarizes it in the interests of clarity.

References cited in the text are given at the end of the appropriate chapter. There is a mix of primary and secondary references: the primary references are given when no suitable secondary reference exists, while secondary references to texts or popular books are given for reasons of their greater accessibility to the majority of readers. The bibliography contained in Appendix B includes full listings of all references cited, along with other suggestions for reading around the topic of cosmology. It is intended that

Structure of This Book

this will help the reader who wishes to undertake further study of this immense and fascinating subject.

2

The Observed Universe

Modern cosmology is firmly based on observations of the present day universe. Cosmological theory is founded on the attempt to explain these observations in a generally consistent manner by the coherent application of accepted physics. This chapter presents a summary of the most important observed properties of the universe. Enough detail is included to give the reader some feel for how these observations first came to be made, and how they were interpreted. It is this catalogue of known properties of the universe which the cosmological theory described in the rest of this book must explain in physical terms.

2.1 Introduction

The universe has always been available for inspection by interested people, but until the invention of the telescope the only means of examination other than pure thought was the naked eye. For a long time, in fact until the late 1940s, the optical telescope was the only powerful instrument with which the universe could be examined closely, and nearly all cosmological studies assumed that the visible universe was all that there was to explain.

With the invention of radio telescopes in the 1940s, followed in rapid succession by telescopes operating in the microwave bands, and then X-ray and gamma-ray telescopes, a picture of the universe began to build up which drew on all the different parts of the electromagnetic spectrum. More recently, devices which detect neutrinos and gravitational waves have been used to examine even more deeply hidden properties of the universe. Each new device has brought the ability to observe previously unsuspected phenomena, and, in the same way, advances in statistical methods, computational analysis and numerical simulation have opened up new frontiers of vision in the way we see and attempt to understand the universe.

The remaining sections of this chapter broadly survey the present state of knowledge gained from observations of the universe at large. These are the data which motivate our theoretical study of cosmology, and they are also the data against which all theoretical models must be tested and sharpened. The development of theoretical explanations – and through theory, understanding – is the task of the rest of the book.

2.2 Existence of Structure

Even with the naked eye it is possible to observe that the universe is structured. The most useful definition of what we mean by structured is that the universe looks different on different scales of size. The easiest examples to see with the naked eye are the sun and planets of our own solar system, and, looking further, stars and small star clusters such as The Pleiades and the one in the Orion nebula. Our own galaxy, the Milky Way, gets its name from the way it appears as a milky stream across the night sky, showing that it is a large group of stars formed into a flattened disc: its appearance is due to the fact that our sun is one of the stars in the disc. A small telescope or good binoculars can be used to observe the nebulae, other milky luminous objects. Many of these turned out, when powerful enough telescopes became available, to contain large numbers of stars, and are therefore other galaxies like ours. Looking at deep sky photographs taken with the most powerful telescopes shows that there are very large numbers of galaxies having different kinds of structure, and most of these galaxies are themselves grouped into larger objects called galaxy clusters. Clusters of galaxies contain anything from a few galaxies to many thousands. During recent decades, it has been realized that even clusters of galaxies are themselves grouped together into superclusters. Such superclusters are so large that they each occupy a significant fraction of the total sky area. The cosmos therefore seems to contain physical structures on scales all the way up to the entire visible portion of our universe.

2.3 Cosmic Expansion

The idea that the universe may be expanding first began to be taken seriously in the early part of the twentieth century. In 1912, Slipher measured the speed with which the Andromeda galaxy, M31, is moving relative to the earth (Hubble 1936). He did this by taking a spectrum of M31 and comparing it to a laboratory spectrum. The shift in frequency of the spectrum of M31 is interpreted as a Doppler shift, which then allows the velocity of M31 relative to the earth to be calculated. Slipher found that the spectrum of M31 is shifted towards the blue relative to the laboratory spectrum, so Andromeda is approaching us, with a velocity estimated to be 300 km/s. Slipher continued his work on spectra of other galaxies, and by 1924 had obtained over 40 spectra. Of these, 90% were shifted towards the red end of the spectrum, leading to the conclusion that the universe was generally expanding.

Hubble extended Slipher's work by developing estimates of distance, so that the recession velocities of other galaxies could be related to how far they were from the Milky Way. Around 1918, Shapley had derived a distance scale which could be used reliably inside our galaxy. This scale was based on a class of stars known as Cepheids. Cepheids are variable stars whose brightness varies periodically between fixed maximum and minimum brightnesses. Shapley was able to show, by surveying Cepheids in the Milky Way, that there is a fixed relationship between the maximum brightness of a Cepheid and the period of its variation. He then calibrated this relationship accurately, so that if one observes a Cepheid for long enough to determine its period, its intrinsic brightness can be accurately calculated. The apparent brightness of the Cepheid then gives a good measure of its distance from earth. Hubble extended the Cepheid distance measure beyond the Milky Way by observing the light curves (curves of brightness against time) of the brightest variable stars in other galaxies. Some of these variable stars were found to have light curves typical of Cepheids, that is with the same relation between period and brightness. Hubble then assumed that these stars were normal Cepheids, and was able to estimate distances from earth to the galaxies that contained the Cepheids.

Hubble then extended his distance scale even further by using the apparent brightness of the brightest star in a galaxy as the indicator of its distance. This assumption was based on the observation that the brightest stars in galaxies appeared to have similar apparent magnitudes. Hubble then calibrated the brightest star distance measure by applying it to nearby galaxies for which he already had distances measured on the basis of the Cepheid distance scale. After over a decade of work, Hubble published his discovery that the apparent recession speed (v_{galaxy}) of distant galaxies determined from Slipher's catalogue of spectral redshifts is directly proportional to their distance (d_{galaxy}):

$$v_{galaxy} = H d_{galaxy}. \tag{2.1}$$

The constant of proportionality, H, is now called *Hubble's constant*. The dimension of, H, is velocity/distance, which in turn has the dimension of 1/time. The corresponding timescale $1/H$ is an estimate of the time for which the universe has been expanding. The value derived for this constant from Hubble's data was

$$H \simeq 500 \, \text{km.s}^{-1}.\text{Mpc}^{-1} \tag{2.2}$$

Hubble's value for H therefore leads to a value for the age of the universe, t_{age}, of

$$t_{age} \simeq \frac{1}{H} \simeq 2 \times 10^9 \, \text{yr}. \tag{2.3}$$

This age, unfortunately, was uncomfortably shorter than the best estimates of the age of the oldest earth rocks – see section 2.4 below.

It took a long time until it was realized that Hubble had made an inadvertent error in estimating H. This was because the distance scale on which the value of H was based had been estimated using Cepheids. What Hubble could not have known was that there are in fact two different types of Cepheids, and that they have different period–brightness relations. The Cepheids which Hubble had used to calibrate his brightest star measure turned out to be of the second type. When Hubble's constant was recalculated from the revised period–luminosity relation by Sandage in the 1950s, the new value was found to be reduced by an order of magnitude. The exact value of H_{now} is still controversial, with published estimates ranging from as low as $20\,\text{km.s}^{-1}.\text{Mpc}^{-1}$ (based on observations using supernovae as distance calibrators) to as high as $120\,\text{km.s}^{-1}.\text{Mpc}^{-1}$ (based on assumptions about the brightness distribution of galaxies). The current best estimates of H seem to cluster in the range

$$H_{now} \simeq 65\text{--}90\,\text{km.s}^{-1}.\text{Mpc}^{-1}, \tag{2.4}$$

although the uncertainty in these estimates may be very large.

2.4 The Age of the Cosmos

The age of the universe has been a subject of debate for a number of centuries, although for a long time the main issue was whether the universe had existed for all of eternity, or had a finite lifespan. Once a clear theory of the evolution of terrestrial lifeforms had been expounded by Darwin (1859), it became widely accepted that the universe was evolving and had a finite age. It also became a critical scientific problem to estimate the earth's age in order to understand how much time had been available for terrestrial lifeforms to evolve. The resulting controversy (reviewed by Gould 1987, Barrow and Tipler 1986) eventually led to the first thorough investigations of the age of the earth, and as a byproduct generated the technique of radioactive dating.

When Hubble first measured the expansion rate of the universe, it was overestimated by about an order of magnitude due to an unexpected systematic error. Accordingly, Hubble's age for the universe of $t_{cosmos} \sim 10^9$ yr was underestimated by the same factor, and ran into severe disagreement with geological estimates of the earth's age, $t_{earth} \simeq 5 \times 10^9$ yr based on radioactive dating of the oldest known terrestrial rocks. Interestingly, one can apply the technique of radioactive dating to our own galaxy by assuming that all the uranium in our galaxy was formed at a time t_g (Weinberg

The Age of the Cosmos

1972). The initial abundance ratio of uranium isotopes would be

$$\left.\frac{U^{235}}{U^{238}}\right|_{init} = 1.65 \pm 0.15 \tag{2.5}$$

The decay rates of these isotopes are accurately known:

$$\lambda(U^{235}) = 0.971 \times 10^{-9} \text{ yr}^{-1}, \tag{2.6}$$

$$\lambda(U^{238}) = 0.154 \times 10^{-9} \text{ yr}^{-1}. \tag{2.7}$$

The abundance ratio is presently known to be

$$\left.\frac{U^{235}}{U^{238}}\right|_{now} = 0.00723 \tag{2.8}$$

Putting this all together, the total age of the Milky Way is

$$t_{now} - t_g = \frac{\ln\left[\frac{U^{235}}{U^{238}}\right]_{init} - \ln\left[\frac{U^{235}}{U^{238}}\right]_{now}}{\lambda(U^{235}) - \lambda(U^{238})}, \tag{2.9}$$

so that the time elapsed since the Milky Way formed is

$$t_{now} - t_g \simeq 6.6 \times 10^9 \text{ yr.} \tag{2.10}$$

The situation now is that the best estimates of the age of the universe come from the structure and material content of globular clusters. Globular clusters are gravitationally bound star groups, typically containing $\sim 10^6$ stars, orbiting a common centre of mass. Because of the element abundances present in these stars, they are believed to represent the earliest population of stars to have formed. Globular clusters are gravitationally stable, and are assumed to have suffered little mixing with other star populations since they were first formed. Precise determination of the ages of globular cluster stars is difficult, and requires a certain amount of theoretical input in the form of stellar lifecycle calculations. However, the globular cluster stars in our own Milky Way and neighbouring galaxies yield a consensus age in the range of about $(1.4 \pm 0.3) \times 10^{10}$ yr (Vandenberg 1983, Iben and Renzini 1984). Clearly the age of globular clusters provides only a lower limit on the age of the universe. However, this age is in generally good agreement with that estimated on the basis of the current measurements of the cosmic expansion rate. Due to the large fractional uncertainty in the present value of the Hubble parameter, apparently caused by its strong sensitivity to the method used to measure it, we are at best justified in saying that the age of the universe most probably lies somewhere in the interval

$$9 \times 10^9 \text{ yr} \leq t_{now} \leq 20 \times 10^9 \text{ yr.} \tag{2.11}$$

The order of magnitude of this value is well agreed upon, and for most purposes in the rest of this book, we shall generally use the estimate $t_{now} \sim 10^{10}$ yr.

2.5 Element Abundances

Spectral observation of stars and galaxies shows that on average the luminous material in the universe consists mainly of hydrogen (H) and helium (He), with a small amount of the heavier elements (all of which are called 'metals' in the astronomical literature). The exact abundances measured vary from object to object, but generally there appears to be about two or three times as much hydrogen as helium by mass (Weinberg 1972). By contrast, the metals contribute less than 1% of the luminous mass.

It was realized in the late 1950s that the abundances of metals could be explained by production through nuclear fusion reactions in stars, especially given the (then newly revised) age of the universe of 10^{10} yr. This gave rise to the hope that all the element abundances could be understood as having arisen through fusion reactions inside stars which were initially composed entirely of hydrogen. However, it was soon recognized that the amount of helium was too large to be explained in this way. Eventually it came to be accepted that the helium abundance required a cosmological explanation. This explanation was developed in the late 1960s and early 1970s (see Peebles 1971, Weinberg 1972).

In recent years, a new observational approach to the helium abundance question has come to the fore. Rather than averaging a large number of helium abundance observations, this approach is to try to identify the objects (principally galaxies or globular clusters) which formed earliest and establish that they have low metal abundances. These objects can then be argued to have formed from material in which hydrogen, helium and metals were present at their primordial abundances. Observation of such objects has yielded very accurate helium abundance mass fractions, although these are subject to the main difficulty that we still have no effective way of averaging these observations to improve their reliability. In general, such observations agree that the mass abundance X of the most common isotope of helium, He^4, is

$$X(He^4) \simeq 23 \pm 2\%. \qquad (2.12)$$

Understanding why He^4 had this primordial abundance is one of the great successes of modern cosmological theory. This is because if there are no reactions to create He^4, then one would expect $X(He^4) = 0\%$, while if the reactions are effective at fusing hydrogen into helium, one would expect $X(He^4) = 100\%$ (Zeldovich and Novikov 1983). The production of helium and the explanation of the primordial helium abundance in terms of physical constants is the subject matter of Chapter 7.

2.6 Cosmic Matter

Since the 1930s it has been known that as well as normal matter, there can also be antimatter, made of antiparticles. For example, the electron has its own antiparticle, the positron, which has the same mass but opposite charge. When a particle collides with its antiparticle, they annihilate each other and release photons. Positrons, antiprotons and other forms of antiparticle are routinely observed in terrestrial particle physics experiments.

It might seem reasonable to assume that the universe could therefore contain equal amounts of matter and antimatter; however, this conflicts with all known observations. Nearly all the scales which are accessible to observation reveal that the universe consists entirely of matter. Within the solar system there is definitely no antimatter because we would see the spectacular amounts of energy released as it undergoes mutual annihilation with the matter which we know to be present. Terrestrial probes have also made contact with other planets without annihilating.

Cosmic rays tell us about larger scales, since they are believed to originate from the entire galaxy, and probably from other nearby galaxies as well. The antimatter fraction of high energy ($> 100\,\text{MeV}$) cosmic rays is small. The cosmic ray antiproton–proton ratio is

$$\frac{\overline{p}}{p} \simeq 3 \times 10^{-4}, \tag{2.13}$$

while the ratio of antihelium nuclei to helium nuclei (α-particles) is even smaller,

$$\frac{\overline{\text{He}^4}}{\text{He}^4} \leq 3 \times 10^{-5}. \tag{2.14}$$

These values are small enough to be consistent with production of antimatter by particle collisions (see Börner 1988).

On even larger scales, the existence of significant quantities of antimatter is ruled out by the absence of the hard γ-ray emission which would be produced when annihilation occurred in the gas distributed throughout galaxy clusters. The presence of such gas is known from its emissions in the X-ray region (see Kolb and Turner 1990). Nearby clusters of galaxies must therefore consist entirely of matter or antimatter. If both matter and antimatter are present in the universe, they must therefore be segregated on scales corresponding to the mass of galaxy clusters. Since large clusters typically contain at least 100 galaxies, this segregation scale corresponds to a mass of at least $\sim 10^{12} M_\odot$, where $M_\odot$ is the mass of our sun, and probably much more. It is difficult to imagine a physical process which could have segregated antimatter from matter on such large scales. This is not in itself a conclusive argument against matter–antimatter segregation, but the evidence that we have is all in favour of, and totally consistent with, the conclusion that the material of the universe consists entirely of matter.

In the late 1970s, the study of particle physics suddenly became directly relevant to the study of cosmology when the cosmic matter–antimatter asymmetry was explained in terms of microphysical processes predicted by unified particle theories. This explanation is described in Chapter 8.

2.7 Background Radiation

The cosmic background radiation was unequivocally detected in the 1960s (Penzias and Wilson 1965, Dicke *et al.* 1965). It had been pointed out in the 1940s that if the universe had begun by expanding and cooling from an early hot phase, then simple timescale estimates based on the age of the universe would mean that there should be a cool relic of the original hot thermal radiation still left in the universe and that this relic radiation would have a temperature of about 5 degrees kelvin (Alpher *et al.* 1948). At that time, it was thought that such a low temperature could not be detected. Penzias and Wilson (1965) made their discovery using what was then the most sensitive microwave antenna in existence. They found that when they calibrated their antenna and observed the sky in regions where there was no detectable galactic emission, they found that a small but significant amount of radiation was still detectable. Although they were only able to measure the amount of radiation at a fixed frequency, the corresponding thermal radiation was found to be consistent with that of a black body at a temperature of about 3K. Since then, measurements of the background radiation temperature in different regions of the spectrum have established that it has the following properties:

1. The spectrum of the background radiation is definitely that of a black body radiating at a temperature, I, close to 3K (Mather *et al.* 1994):

$$T \simeq 2.726 \pm 0.010 \, \text{K}. \tag{2.15}$$

2. The temperature distribution across the sky contains a hot spot corresponding to the motion of the earth in the direction of the Virgo supercluster at a velocity (Davies 1988)

$$v_{earth} \simeq 360 \pm 30 \, \text{km/s}. \tag{2.16}$$

3. The radiation is highly uniform. The degree of fractional variation of the temperature averaged over the whole sky is (Mather *et al.* 1990)

$$\frac{\Delta T}{T} \simeq 10^{-5}. \tag{2.17}$$

The first of these facts means that the radiation is in thermal equilibrium, suggesting that it is of primordial origin and not the product of other cosmological events like galaxy formation. This is confirmed by the third point, since it is unlikely that other physical processes that produced significant amounts of radiation would have been so effectively thermalized by now that there would be no trace distortion left in the spectrum of the background radiation. Once the effect of the velocity deduced from the second fact has been compensated for, the radiation temperature is found to be independent of direction. The even distribution of the average temperature over the sky is therefore a strong argument in favour of isotropy – see section 2.9 – and the smallness of the temperature variation likewise strongly favours the interpretation that the primordial universe was homogeneous, that is to say, it was the same everywhere.

2.8 Mass Density

The amount of mass represented by the visible luminous material in the universe can be estimated by deducing values for the mass to luminosity ratio, M/L, of large scale systems from galaxies to superclusters. The mass density of the universe is usually expressed in terms of the density parameter, Ω, which is the ratio between the measured mass density and the critical density, ρ_c. The critical density is calculated theoretically and corresponds to the density required to make the cosmic expansion rate just fast enough that the expansion will not be halted by the gravity of the matter. The evolution of the density parameter provides the material for Chapter 5, where these relationships will be explored in detail.

Estimates of M/L appropriate to averaged samples of galaxies give the result that

$$\Omega_{galaxy} \simeq (1\text{–}2) \times 10^{-2}. \tag{2.18}$$

This value can be changed by approximately an order of magnitude if one examines larger scale structures in the same way, although the uncertainties are also increased since M/L is harder to estimate on larger scales.

Alternative ways to estimate Ω are to extrapolate the mass density of the Virgo supercluster to the whole universe. The mass of the Virgo supercluster can be deduced from the velocity (2.16) with which the earth is now moving towards it. The resulting estimate for the density parameter is (Börner 1988)

$$\Omega_{supercluster} \simeq (9 \pm 4) \times 10^{-2}. \tag{2.19}$$

Note that this estimate requires input from determination of the age of the universe and the cosmic background radiation temperature measurements, but is independent of estimates of M/L. It is also significantly larger than that predicted on the basis of the direct measurements of luminous material.

Theoretical models relating the luminosity and density of infrared sources to the rest of the matter distribution allow the IRAS (Infrared Astronomical Satellite) data to be used to derive a value for Ω. Rowan-Robinson et al. (1986) found the value

$$\Omega_{IRAS} \simeq 0.85 \pm 0.16. \tag{2.20}$$

This value is larger again than either of the previous values. Two conclusions must be drawn from the measurements of the mean mass density. Firstly, the density parameter is at least as large as 0.01, since this is derived on the basis of direct observation. Secondly, there may be a significant amount of invisible material in the universe. If so, then it will have important consequences for cosmological theory, since it will contribute a large share of the gravitation, and gravity is what theory tells us drives the evolution of the cosmos. The questions relating to the possible existence of dark matter are examined further in Chapter 11.

2.9 Isotropy and Homogeneity

Observations of all kinds on large scales reveal that although the universe contains structure on all scales, the kind of structure seen is consistent with the universe being isotropic and homogeneous. The statement that the universe is isotropic means that what is observed has no dependence on direction. The evidence that our universe is isotropic is founded on a number of different observational facts (see section 2.7).

Hubble's law

The cosmic expansion discussed in section 2.3 is found to be independent of direction to a high degree. This is established by surveying the nonradial components (peculiar velocities) of the motion of distant galaxies and provides evidence for homogeneity as well as isotropy (Peebles 1971).

The galaxy distribution

The detailed galaxy number count survey carried out by Hubble (described in detail in Hubble 1936) showed that as well as being isotropic to a high degree, the galaxy distribution has the same form all the way out to a distance of about 1000 Mpc. The absence of any downturn in the galaxy number count demonstrates that up to these scales, the universe is homogeneous and has no edge. This must be interpreted as strong evidence for homogeneity as well as isotropy (Peebles 1971, Rowan-Robinson 1977).

Clusters of galaxies

The sky catalogues of galaxy clusters compiled by Shapley and Ames (1932) and Abell (1958) show that galaxy clusters are evenly distributed across the sky. The evidence is discussed by Peebles (1971) and Raine (1981).

Evidence for Evolution

Radio and infrared sources

Radio source objects have been thoroughly surveyed in the region of the sky away from the plane of the galaxy. Their distribution is found to be isotropic to within less than 10% (Rowan-Robinson 1977). Likewise, thousands of infrared sources were surveyed by the Infrared Astronomical Satellite (IRAS), with the conclusion that they are even more smoothly distributed than the luminous galaxies (Rowan-Robinson *et al.* 1986).

The background radiation

The cosmic background radiation temperature is also found to be independent of sky direction, and is completely smooth across the sky to one part in 100 000. This may be thought to be due to the effects of recent scattering, but if this is the explanation then the scattering medium would have to be something which scatters strongly across precisely the waveband surveyed by the Cosmic Background Explorer (COBE) satellite, while being completely transparent in the radio and visible regions of the spectrum.

The evidence is therefore clearly in favour of the interpretation that the universe is isotropic and homogeneous. Furthermore, the observations listed here can be seen as providing estimates of the variation of the gravitational potential through the universe. Calculation of this variation shows that the gravitational potential of the universe is homogeneous and isotropic to the same accuracy as the background radiation: the variation must be less than one part in 100 000 (Barrow 1989). It is this high degree of smoothness in the gravitational structure of the universe that must be accounted for by any effective theory of cosmology.

2.10 Evidence for Evolution

Until the early twentieth century, the general view of the cosmos was that it was stationary, unchanging, immutable, and that this was so because there was no other way in which physical laws could be expected to operate reliably. The work of American astronomers in the first third of the century provided the evidence which gave cosmologists the impetus to challenge these assumptions (Bertotti *et al.* 1990).

There are now obviously sound theoretical reasons for believing that the universe is evolving. However, direct evidence that complex objects such as galaxies did not always exist, but formed during a particular cosmic epoch and have evolved since then, first came in the 1950s with work on the number counts of radio source galaxies and then, in the 1960s, with results on the number density of quasars. These results are superbly surveyed in the classic book by Sciama (1971). The main results are that, in the case of the radio sources, their counts drop off more steeply with distance than would be expected if they were distributed uniformly through sky depth. This would appear to be in contradiction with homogeneity, which is evinced by other data as we have seen in section 2.9. The evolutionary interpretation of the radio source counts is therefore supported: the number counts drop off steeply because as we look further out in space, we also look back in time, and when we look back far enough, we see the time when radio sources were still forming, so that there were fewer to see than there are now. The universe must therefore have evolved during the last few billion years.

The later work on quasar number counts shows a similar effect: beyond a certain distance corresponding to looking back in time a few billion years, the number density of quasars falls off more rapidly than would be expected if quasars had always existed. In fact, it is now generally believed that quasars as well as radio sources reside in galaxies, and that quasars are probably earlier stages in the evolution of galaxies.

Finally, the cosmic background radiation also provides evidence that the universe is evolving: the temperature of the background radiation was higher in the past. Measurements of the excitation spectra of neutral carbon in a gas cloud associated with the distant quasar Q1331+170 yield a temperature for the background radiation of 7.4 ± 0.8 K (Songaila *et al.* 1994). This value is consistent with the idea that the background radiation would have cooled as the universe expanded.

These results confirm the idea that the universe has evolved to its present state. Cosmological theory must therefore explain not only how the universe can exist in its present state, but how it can evolve into it from earlier states. Explaining the universe requires a dynamical theory of cosmology.

2.11 Causal Horizons

The question of causality is interesting because it is quite easy to state, and yet requires a very subtle explanation which can only be comprehended after nearly reaching the end of this book, where the solution is given in Chapter 9. The problem arises as follows:

We know that physical explanation is only useful when causality is respected. That is to say, when some physical fact is observed, there must be some cause of that fact. The cosmological causality problem begins in the observation that if one holds one's arm outstretched and holds up two fingers to the sky, the deep regions of sky visible on either side of the fingers are separated by a distance of more than 10 billion light years (Clutton-Brock 1993). Thus, the regions of cosmic background radiation reaching one's eyes from different sides of one's hand can never have been in physical contact with each other – they are separated by a distance greater than that which could have been travelled by light in the age of the universe. How could those two regions have been caused to be at the same temperature and density as each other to such a high accuracy? This question was first posed by Rindler (1959) and is now known as the *horizon problem*. The full force of this problem was first appreciated when it was expounded by Dicke and Peebles (1979). Although simple to state, its solution is an essential aspect of a causal physical theory of the universe.

2.12 Review

The quantity of cosmological observations forces any discussion of the observed properties of the cosmos to be simplified in the interests of brevity. Since this book is primarily about the physics of cosmology, observations are mainly confined to those presented here. There are many books and review articles which present far more detail than is possible here. The interested reader should consult Sciama (1971) and Silk (1980) for extremely readable accounts. Both Peebles (1971) and Weinberg (1972) give detailed technical descriptions of a multitude of observations and their interpretation. The books by Börner (1988) and Kolb and Turner (1990) contain thorough surveys of the subject and valuable updates to earlier books. Much of the thought and many of the discoveries described in this chapter have been related by the people who were directly involved. A rich vein of those writings is to be found in Bertotti *et al.* (1990). The main observed properties of the universe which must be taken into account in constructing a cosmological theory are

1. The existence of structures on all length scales;

2. Isotropy and homogeneity on galaxy cluster scales;

3. Expansion according to Hubble's law (equation 2.1);

4. A cosmic age of the order of about 10 billion years;

5. The helium fraction of about 1/4 of the luminous mass;

6. Matter as the only component of the cosmic material;

7. Mass density within an order of magnitude of closure density;

8. Smooth thermal radiation background at 3 degrees Kelvin;

9. Evidence for cosmic evolution from radio sources and quasars;

10. Apparent lack of causal connection between similar regions.

The rest of this book is concerned with developing the dynamical, physical, causal, cosmological theory which will account for the observed properties of the cosmos.

2.13 References

Abell, G. *Astrophys. J. Suppl.* **3**, 211 (1958).

Alpher, R.A., Bethe, H.A. and Gamow, G. *Phys. Rev.* **73**, 803 (1948).

Barrow, J.D. *Quart. J. Roy. Astron. Soc.* **30**, 163 (1989).

Barrow, J.D. and Tipler, F.J. *The Anthropic Cosmological Principle* (Oxford University Press, 1986).

Bertotti, B., Balbinot, R., Bergia, S. and Messina, A. *Modern Cosmology in Retrospect* (Cambridge University Press, 1990).

Börner, G. *The Early Universe: Facts and Fiction* (Springer-Verlag, Berlin, 1988).

Clutton-Brock, M. *Quart. J. Roy. Astron. Soc.* **34**, 411 (1993).

Darwin, C. *On the Origin of Species by Means of Natural Selection* (John Murray, London, 1859).

Davies, R.D. *Quart. J. Roy. Astron. Soc.* **29**, 443 (1988).

Dicke, R.H. and Peebles, P.J.E. in Hawking, S.W. and Israel, W. (eds) *General Relativity: An Einstein Centenary Survey* (Cambridge University Press, 1979).

Dicke, R.H., Peebles, P.J.E., Roll, P.G. and Wilkinson, D.T. *Astrophys. J.* **142**, 414 (1965).

Gould, S.J. *Time's Arrow, Time's Cycle* (Harvard University Press, Cambridge, Massachussetts, 1987).

Hubble, E.P. *The Realm of the Nebulae* (Yale University Press, 1936).

Iben, I. and Renzini, A. *Phys. Rep.* **105**, 331 (1984).

Kolb, E.W. and Turner, M.S. *The Early Universe* (Addison-Wesley, New York, 1990).

Mather, J.C. *et al. Astrophys. J.* **354**, L37 (1990).

Mather, J.C. *et al. Astrophys. J.* **420**, 439 (1994).

Peebles, P.J.E. *Physical Cosmology* (Princeton University Press, 1971).

Penzias, A.A. and Wilson, R.W. *Astrophys. J.* **142**, 419 (1965).

Raine, D.J. *The Isotropic Universe* (Adam Hilger, Bristol, 1981).

Rindler, W. *Mon. Not. Roy. Astr. Soc.* **116**, 662 (1956).

Rowan-Robinson, M. *Cosmology* (Clarendon Press, Oxford, 1977).

Rowan-Robinson, M., Walker, D. and Yahil, A. *Astrophys. J.* **301**, L1 (1986).

Sciama, D.W. *Modern Cosmology* (Cambridge Unversity Press, 1971).

Shapley, H. and Ames, A. *Ann. Harvard College Obs.* **88**, 41 (1932).

Silk, J. *The Big Bang* (W.H. Freeman, San Francisco, 1980).

References

Songaila, A. *et al. Nature* **371**, 43 (1994).

Vandenberg, D.A. *Astrophys. J. Suppl. Series* **51**, 24 (1983).

Weinberg, S. *Gravitation and Cosmology* (Wiley, New York, 1972).

Zeldovich, Y.B. and Novikov, I.D. *Structure and Evolution of the Universe* (University of Chicago Press, 1983).

3

The Cosmological Equations

This chapter is concerned with the derivation of the equations that describe the evolution of the universe. These equations are a direct consequence of the need for the universe to conserve energy in each local volume of space. Hubble's expansion law can be easily understood from the predicted form of the expansion equations. Some of the simplest solutions to the expansion equations are presented and used to throw light on other cosmological questions.

3.1 The Dynamical Equations from Cosmic Energy

The universe appears, to us, to be isotropic about our own position in it – that is, it is generally the same in all directions. From this, we must deduce *either* that we occupy the central position in the cosmos, *or* that the universe appears isotropic about every point. The latter assumption, which explains our observation of the spherical symmetry of the cosmos, is known both as the *Copernican principle* and as the *cosmological principle*. It is in accord with the scientific principle known as *Occam's razor*, which states that one should make no unnecessary assumptions. In other words, one should work from the minimum set of hypotheses required to make a workable mental model of the observed phenomena.

In order to derive the equations governing the large-scale behaviour of the universe, we shall require only this principle and the well known relativistic relation

$$E = mc^2, \qquad (3.1)$$

where c is the speed of light, which states the equivalence between mass, m, and energy, E. From this point on, we shall use units in which $c = 1$ identically. (See Appendix A for a detailed description of these units. The missing factors of c are easily restored at the end of the calculation, if this should prove necessary.) The cosmological equations can now be derived as follows. Consider an infinite, homogeneous gas cloud, with mass density ρ.

Now choose an arbitrary origin at a point A – arbitrary because the cloud is homogeneous – and imagine a thin spherical shell centred on A, with radius r. When we think of the gas cloud as infinite, we mean, practically speaking, that it should be much bigger than the actual shell whose movements we will follow. This is obviously necessary to keep our analysis free of the problems arising from boundary effects. 'Infinite' therefore means the same as 'boundary so far distant as to be irrelevant' for our present purposes.

We want to find the equations describing the motion of the thin shell, because such equations will govern the motion of any and every spherical shell centred on any point. This is true precisely because we are dealing with a universe which is isotropic about every point.

The first important consideration is the energy balance at each point on the gas shell: the first relevant quantity is the kinetic energy per unit mass,

$$T = \frac{1}{2}v^2 = \frac{1}{2}\dot{r}^2. \tag{3.2}$$

The potential energy is a little trickier, but can be found by use of two results originally obtained by Newton. The first is that no net gravitational force is exerted by matter exterior to a spherical shell on matter inside the shell (this result is usually called Newton's *iron ball theorem* – the result should also be familiar from electrostatics). The consequence of this in our case is that we can entirely ignore the gas outside the shell which we have chosen to consider. The second is that the gravitational field outside any homogeneous sphere is the same as that produced by a mass point having the same mass as that of the sphere, and located at the centre of the sphere. So the ball of material located inside the shell we are considering directly affects the motion of the shell only through its mass and its radius. The mass is clearly

$$M = \frac{4}{3}\pi r^3 \rho, \tag{3.3}$$

so the potential energy per unit mass is

$$V = -\frac{GM}{r} = -\frac{4\pi G}{3}r^2 \rho \tag{3.4}$$

(where G is Newton's constant of gravitation.) Now the principle of energy conservation requires that the sum of these two energies must not change:

$$T + V = \text{constant}, \tag{3.5}$$

because any flow of energy into the shell from one direction must be balanced by an equal and opposite flow – otherwise the isotropy would be broken, since there would be a preferred direction, namely the direction of the net energy flow. The energy balance equation for the gas shell can thus be written as

$$3\left(\frac{\dot{r}}{r}\right)^2 = 8\pi G \rho - 3K, \tag{3.6}$$

where it is easy to see that $K = \kappa/r^2$ with κ an as yet undetermined constant. The sign of the K term has been chosen in accord with common usage, and its role will become clear in the course of this chapter. It is important to note that there is nothing, at least in principle, that restricts the sign or value of the constant κ.

Clearly, equation (3.6) only suffices to specify $r(t)$ for the gas shell if we know the form of $\rho(t)$. It should also be clear that if the shell expands or contracts, then so must the sphere it encloses, or else the cloud will lose its homogeneity. Normally, we do not know $\rho(t)$, and so we need another equation to specify the solution. The necessary relation is provided by the first law of thermodynamics, which relates the change in energy E to the changes in entropy Σ and volume V through the temperature T and pressure P:

$$dE = T\,d\Sigma - P\,dV. \tag{3.7}$$

The simplest case to treat is that with the entropy constant, which corresponds to the expansion being reversible. Assume that this is so: then since $c = 1$ and the main contribution to the energy comes from the mass of the gas, we have $E = \rho r^3$ by (3.1), so

$$d(\rho r^3) + P d(r^3) = 0, \tag{3.8}$$

and hence

$$\frac{d\rho}{dt} + \frac{3}{r}\frac{dr}{dt}(\rho + P) = 0. \tag{3.9}$$

Equation (3.9) is a separable differential equation, so if we have an *equation of state*, that is, a relation of the form

$$P = P(\rho), \tag{3.10}$$

then equation (3.9) has a solution $\rho = \rho(r)$, which can be substituted into equation (3.6). This will then yield – at least in principle – a solution $r = r(t)$.

Equation (3.6) is an equation for the energy balance, but together with equation (3.9) it gives an equation for the acceleration of the gas shell: taking the time derivative of (3.6),

$$6\frac{\dot r}{r}\left(\frac{\ddot r}{r} - \frac{\dot r^2}{r^2}\right) = 8\pi G\dot\rho - 3\dot K \tag{3.11}$$

and substituting from (3.6) for $(\dot r/r)^2$ and from (3.9) for $\dot\rho$ and then rearranging, we find

$$3\frac{\ddot r}{r} + 4\pi G(\rho + 3P) = 0, \tag{3.12}$$

where we have also used the fact that $K \sim 1/r^2$. Equation (3.12) shows that the gas pressure increases the gravitational force contribution due to the mass density. So increasing the internal pressure of the gas cloud will increase the restraining gravitational force which it exerts upon its own expansion. At first sight, this is somewhat surprising, since in other physical situations one studies one finds that internal pressure tends to support the gas against external forces. That this does not work against the gas's own weight is due to the remarkable properties of the gravitational force, mainly the fact that it is always attractive. Physically, the gravitationally attractive nature of internal pressure makes perfectly good sense: in order to increase the internal pressure of a gas, it must have its internal thermal energy increased, and this thermal energy in its turn contributes to the gravitational mass through mass–energy equivalence. Note also at this stage that (3.12), being a second-order differential equation, is the equation describing the dynamical behaviour of $r(t)$. Physically, however, it is conceptually simpler to think of the expansion as a consequence of the energy conservation relations (3.5) and (3.7).

Equations (3.6),(3.9) and (3.12) are the fundamental equations that, together with an equation of state (3.10), completely specify the solution for the motion of the gas cloud. Of course, only two of these three equations are independent, since we derived (3.12) from the other two. Some remarks are in order here:

- We assumed the gas cloud was homogeneous as well as isotropic. The equations we have derived will therefore be applicable to any gas shell centred on any point. We can clarify this by choosing the coordinates to be *comoving* with the gas particles: choose an arbitrary origin, and label each gas particle at some initial time by the (vector) coordinate $\vec{r}_0$ = constant. The motion of the particles can then be described in terms of a scale factor S, where $\vec{r}(t) = S(t)\vec{r}_0$ is the particle position at any time t. S has no spatial dependence because of the isotropy and homogeneity. The dynamical relations can thus be written as equations for the evolution of the scale factor $S(t)$:

$$3\left(\frac{\dot{S}}{S}\right)^2 = 8\pi G\rho - \frac{3k}{S^2}; \qquad (3.13)$$

$$3\frac{\ddot{S}}{S} + 4\pi G(\rho + 3P) = 0; \qquad (3.14)$$

$$\dot{\rho} + 3\frac{\dot{S}}{S}(\rho + P) = 0. \qquad (3.15)$$

Here $k \equiv \kappa/r_0^2$. Notice that this means k is a constant, having the same value for all shell radii, since κ depends on $\vec{r}_0$, $\kappa \sim r_0^2$. This is in keeping with the original assumption of isotropy and homogeneity.

- At no point in the derivation of these equations have we made use of any relativistic relations besides equation (3.1). Nevertheless, equations (3.13–3.15) are precisely the same as those obtained by use of the general relativistic field equations in the case of a spatially homogeneous and isotropic universe. The equations (3.13–3.15) were in fact originally obtained using relativistic methods by Friedmann (1922, 1924). They were first derived using arguments similar to those presented here a decade later by Milne (1934), and McCrea and Milne (1934). The coincidence of the results is not at all surprising, inasmuch as the Einsteinian and Newtonian theories have all the ingredients which were used in deriving (3.13–3.15) in common: in both theories, the gravitational potential $\Phi \sim 1/r$, and here the main contribution to the relativistic energy density is supplied by the rest mass of the gas particles.

What, now, is the application of equations (3.13–3.15)? After all, we have called them the *cosmological equations*. The answer to this question is straightforward, if lengthy. Think of the galaxies in the universe as being the gas particles. The number of galaxies contained in the visible universe is sufficiently large that this approximation can be expected to hold. Then equations (3.13–3.15) describe the time evolution of the mass density function ρ and the *cosmic scale factor S*. The interpretation of S is also simple: label each galaxy by a unique coordinate $\vec{r}_0$ centred on our own galaxy. We are free, by the homogeneity and isotropy, to choose any centre, and this choice is the obvious one because it makes the isotropy about our position explicit in the form of the equations. Then the cosmological equations tell us how the other galaxies move relative to ours, by giving the evolution of $\vec{r}(t) = S(t)\vec{r}_0$. Note that this procedure could, in principle, be carried out unchanged by any sentient observer anywhere in the universe.

3.2 Derivation and Explanation of Hubble's Law

The equations derived so far allow us to produce an explanation of the linear velocity–distance relation found by Hubble. Recall that this was of the form

$$v = H_0 r \tag{3.16}$$

where v is the recession velocity of a galaxy at a radial distance r, and H_0 is the parameter now known as the Hubble constant. Equation (3.16), as we will see here, is the expected form of the velocity–distance relation in a homogeneous and isotropic universe. Recalling that we have agreed to treat the galaxies as being the gas particles, consider the motion of a gas shell at radius r, centred on our position, and expanding outwards. The velocity of the shell is, by the preceding argument:

$$v = \dot{r} = \left(\frac{\dot{r}}{r}\right) r = \left(\frac{\dot{S}}{S}\right) r. \tag{3.17}$$

We showed earlier that $\dot{S}/S$ is a function only of time t, and is, therefore, spatially constant. Accordingly, we can make the identification

$$H \equiv \frac{\dot{S}}{S}, \tag{3.18}$$

and then identify H_0 as the present day value of $H(t)$, bringing agreement with Hubble's result. Note that $H = H(t)$ is still allowed to vary with time. Hubble's original results were not sensitive to this variation, since his observations were restricted to galaxies relatively close to our own. As one looks out further into space, however, one sees galaxies at earlier times in the history of the universe, because of the fixed speed of light, c. The variation in time of the Hubble parameter then becomes crucial to interpreting the observational data. This point was discussed further in Chapter 2 on the observed properties of the universe. Note also that the homogeneous and isotropic universes all reproduce the Hubble relation (3.16), regardless of the values of H, ρ, P, or k.

3.3 Solutions to the Expansion Equations

It is possible to derive a large volume of information from the cosmological equations (3.13–3.15) without actually having to solve them in any great generality. Firstly, we can simplify them for use in the present section by neglecting the pressure terms. This should provide a reasonable approximation to the expansion of the universe, at least at the present time, because we have treated the galaxies as gas particles, and we know that galaxies are, in the main, noninteracting. That is to say, they do not often experience direct collisions, so that they behave more like particles in a dust cloud than in a gas. In this case, the pressure terms are effectively negligible, and the equations read

$$3H^2 = 8\pi G\rho - \frac{3k}{S^2}, \tag{3.19}$$

$$3\dot{H} + 3H^2 + 4\pi G\rho = 0, \tag{3.20}$$

$$\dot{\rho} + 3H\rho = 0. \tag{3.21}$$

It should be noted here that the thermodynamic relation (3.15) has here become equation (3.21), which expresses merely the conservation of mass. To see this, notice that from the definition of H, we have $\dot{\rho}/\rho = -3\dot{S}/S$, which can be written $d(\ln \rho)/dt = -3d(\ln S)/dt$ and then integrated to give $\rho/\rho_0 = (S_0/S)^3$. In other words, $\rho \sim 1/S^3$, which is just the expression of the fact of conservation of mass in the expansion of the universe. It is always useful to perform such checks in order to see that the results we derive from calculation are in good agreement with those derived from common sense. Now equation (3.20) can be written as

$$3\ddot{S} = -4\pi G\rho S, \tag{3.22}$$

Solutions to the Expansion Equations

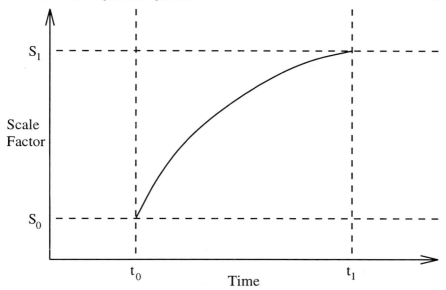

Figure 3.1 *Cosmic scale factor evolution from t_0 to t_1.*

so it is clear that the rate of expansion is decreasing with time. Alternatively, observe that from (3.20) we have $\dot{H} < 0$ always. If we draw the graph of $S(t)$, then the slope will decrease with time, so that if $S = S_0$ and $H = H_0$ at time t_0, then at a later time t_1, we will have $S_1 > S_0$, and $H_1 < H_0$. The graph of $S(t)$ from t_0 to t_1 will thus look like a smoothly increasing function whose slope (first derivative) is smoothly decreasing. The graph of $S(t)$ over the interval from t_0 to t_1 will then look something like that shown in Figure 3.1.

The next obvious question to ask is whether H can ever become zero. Clearly, if $k \leq 0$, then H stays positive, but it can become zero if $k > 0$. Suppose that at t_0, $H = H_0 > 0$ and $\rho = \rho_0 > 0$, with $k > 0$. For $H_0 > 0$, we must have $8\pi G \rho_0 > 3k/S_0^2$. But for the case $P = 0$ as examined in this section, we found that $\rho \sim 1/S^3$, and so, in the course of the expansion, eventually ρ will have decreased until $\rho = 3k/8\pi G S^2$. At this point in time, we will have $H = 0$.

What happens after H becomes zero? Equations (3.19–3.20) tell us only that $d\rho/dt$ goes through zero when $H = 0$. But we can infer the answer to our question without direct calculation: the Newtonian laws from which equations (3.19–3.20) were originally derived are invariant under time-reversal (which is to say that they remain the same if we change $+t$ to $-t$ throughout), so we need only to think of the point at which $H = 0$ as a new initial point of the equations. From this point, the mutual gravitational

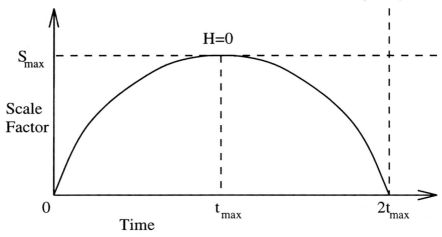

Figure 3.2 *Cosmic scale factor evolution through the time when $H = 0$. In these models, the Hubble parameter H eventually becomes zero, after which the universe recollapses. Since the collapse phase is symmetric with the expansion phase, the total time for which the universe lasts is twice that required for it to reach its maximum size.*

attraction of the dust particles will cause every volume to collapse, with this collapse being the exact time-reverse of the expansion which it had initially undergone. For $k > 0$, the evolution curve will first rise smoothly from its initial value, with its slope steadily decreasing to zero, when the curve reaches its maximum. After that, the curve decreases smoothly with its slope continuing to decrease steadily. This behaviour is depicted in Figure 3.2.

The next question which we can ask is what the past of each of these models looks like. So far, we have taken the initial conditions as being given at some time t_0. If we take this to be the present time, then we can ask how the universe evolved through its past, into the state in which we find it at the present. Again, we can set initial conditions at t_0, and then think of evolving the universe backwards in time. The size of any volume then shrinks, so that ρ increases. Since $\rho \sim 1/S^3$, the k term becomes negligible, for S smaller than some value, say, S_c. Then $3H^2 \simeq 8\pi G\rho$, and as ρ increases, so does H, which means that S shrinks more rapidly as it decreases. From equation (3.22), we have

$$\ddot{S} = -\frac{4\pi G}{3} \frac{S_0^3}{\rho_0 S^2} \sim -\frac{1}{S^2}, \qquad (3.23)$$

which makes the conclusion clear. This is just an expressive way of saying that the universe is, under reasonable conditions, *gravitationally unstable*.

Solutions to the Expansion Equations

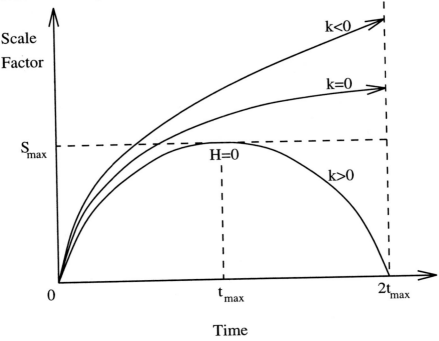

Figure 3.3 *Cosmic scale factor evolution for all signs of k. The case k > 0 is like that shown in Figure 3.2, which is the only case where eventually the Hubble parameter H becomes zero. In the cases for which $k \leq 0$, the expansion never stops, with the k = 0 models expanding only just fast enough to avoid recollapsing.*

There is nothing, at least in the equations derived above, to prevent this runaway collapse from continuing all the way to $S = 0$, where $\rho \to \infty$. This means that our universe started from a *singular point*. This observation forms the basis of all the general relativistic results known collectively as the *singularity theorems*.

The arguments developed above allow us to complete the evolution diagrams shown so that they include the point $t = 0$, and all signs of k can be shown on the same diagram. (See also the exercises at the end of this chapter, and also the discussion given in the book by Weinberg (1972).) The evolution curves for the different signs of k are shown in Figure 3.3.

We can easily deduce that models with $k > 0$ recollapse in a finite time t_r. By the argument already given, if t_{max} is the time at which S attains its maximum, then the recollapse time is $t_r = 2t_{max}$. Models with $k < 0$, on the other hand, expand forever, while models with $k = 0$ only just manage to expand forever. In deriving the evolution equations (3.13–3.15), we used the usual energy equations, and the results that have been deduced from these equations are in accordance with our usual conceptions of energy and the *escape to infinity*. Models with $k > 0$ correspond to test particles being gravitationally bound, so that they always fall back, $k = 0$ models have test particles with zero net energy, so that they can only just escape to infinity (where they have no kinetic energy left), and $k < 0$ models have test particles whose velocity is always greater than the escape velocity, so that they escape to infinity where they still have positive kinetic energy. The simple analogy we originally chose to use for reasoning about the behaviour of the dust particles in the universe has thus been shown to be very powerful.

3.4 Review

The expansion of the universe can be understood as a simple consequence of the laws of energy conservation once account is taken of the energy stored up in the mass of matter. The form of the expansion equations shows that the universe is expanding away from an initially dense phase, and that, in fact, it started from a state with zero volume and infinite density.

Depending on the sign of the energy term in the evolution equations, the universe will either continue to expand forever, or may eventually stop its expansion, turn around and collapse again. The dividing case is the one in which each particle of matter has exactly the escape velocity required to expand away to infinity from the gravitational influence of the sphere of matter on whose surface it lies.

3.5 Exercises

Exercise 3.1 *In the derivation of the cosmological equations, the rotation of matter was ignored. If the matter were rotating, the kinetic energy per unit mass would have an extra contribution like*

$$\frac{1}{2}r^2\omega^2$$

where ω is the shell's angular velocity. Derive the correspondingly modified cosmological equations and study their solution properties.

Exercise 3.2 *Assuming that the pressure of matter can be neglected, plot the evolution curves showing the cosmic scale factor S as a function of cosmic time t for each of the cases $k < 0$, $k = 0$, and $k > 0$. In each case, derive an expression for the age of the universe in terms of the present values of the first and second derivatives of the cosmic scale factor.*

Exercise 3.3 *Assuming that the energy density is dominated by radiation with the equation of state $P = \rho/3$ and that the energy density and pressure of matter can be neglected, plot the evolution curves showing the cosmic scale factor S as a function of cosmic time t for each of the cases $k < 0$,*

$k = 0$, and $k > 0$. In each case, derive an expression for the age of the universe in terms of the present values of the first and second derivatives of the cosmic scale factor. Compare the results with those found for Exercise 3.2.

3.6 References

Friedmann, A.A. *Z. Phys.* **10**, 377 (1922).

Friedmann, A.A. *Z. Phys.* **21**, 326 (1924).

McCrea, W.H. and Milne, E.A. *Quart. J. Math. (Oxford)* **5**, 73 (1934).

Milne, E.A. *Quart. J. Math. (Oxford)* **5**, 64 (1934).

Weinberg, S. *Gravitation and Cosmology* (Wiley, New York, 1972).

4

Cosmological Redshift and Horizons

The cosmological expansion equations allow interesting inferences to be drawn about the observations which we can now make of cosmological objects. The redshifting of light observed from distant objects can be understood in terms of the effect of the expansion on the journey of a photon from a distant object. Furthermore, evaluating how far a photon could have travelled in the time since the expansion of the universe began gives rise to the cosmologically crucial concept of horizons. Wherever horizons exist, they place limitations upon what aspects of the universe and its evolution are directly observable at each place and time in the universe.

4.1 Theory of the Redshift

We now proceed to demonstrate the relationship between the redshift of a cosmological object such as a galaxy or a quasar, and the comoving size of the universe. The importance of this result is both observational and theoretical. This is because it relates the redshift of a cosmological object, which is a directly measurable quantity, to the expansion rate of the universe through its history. The picture we have in mind is one of a galaxy E emitting a light signal at a time t_e, which is then received at a second galaxy R at a time t_r – of course, we are thinking specifically of our galaxy as being placed at R. Also, we assume that the spatial separation between the galaxies is much larger than their own spatial extents, so that they can be treated essentially as points.

If we recall that for a light signal, the velocity, $dx/dt = c = 1$, and that by the definition of our comoving coordinates $dx = S\,dr$, then we find that the coordinate distance between E and R is

$$r_{ER} = \int_{t_e}^{t_r} dr = \int_{t_e}^{t_r} \frac{dt}{S}. \qquad (4.1)$$

Now consider another signal, emitted soon after the previous one, and note

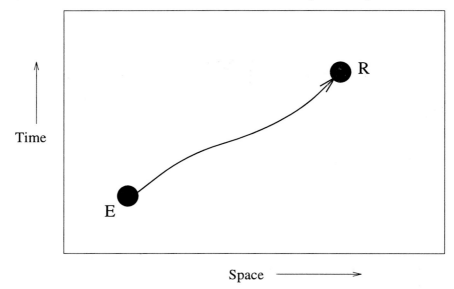

Figure 4.1 *Transmission of a light signal by E and its reception by R. Spatial separation is measured horizontally, time separation vertically.*

that r_{ER} does not change if the times of emission and reception are both slightly shifted by small delays, δt_e and δt_r say. Then we have an alternative expression for r_{ER}:

$$r_{ER} = \int_{t_e+\delta t_e}^{t_r+\delta t_r} \frac{dt}{S}. \qquad (4.2)$$

Equating the right hand sides of equations (4.1) and (4.2), we find

$$\int_{t_e}^{t_e+\delta t_e} \frac{dt}{S} = \int_{t_r}^{t_r+\delta t_r} \frac{dt}{S}. \qquad (4.3)$$

Now we have taken both δt_e and δt_r to be small, so that $S(t)$ does not change appreciably over the respective ranges of the integrations in equation (4.3). The integrals can thus be well approximated by use of the mean value theorem, which tells us that

$$\frac{\delta t_e}{S(t_e)} = \frac{\delta t_r}{S(t_r)}. \qquad (4.4)$$

The point of deriving this last relation is that we can apply it to a conceptually very special case, namely that in which the two signals we consider are chosen to be successive wavecrests of a ray of light of wavelength λ. Remembering that the local speed of light is always constant at $c = 1$, we find that the difference between the times of emission and reception of each of the wavecrests corresponds also to there being a change in the wavelength of the received light signal compared to that of the emitted signal. This is given by

$$\frac{\delta t_r}{\delta t_e} = \frac{\lambda_r}{\lambda_e} = 1 + \frac{\lambda_r - \lambda_e}{\lambda_e} = 1 + Z. \tag{4.5}$$

The last part of equation (4.5) defines the *redshift*, Z, of a light ray. Note that (4.5) is completely general in the sense that it contains no information about the cause of the redshift being due to the expansion of the universe: this is the specific case to which we shall apply it. This equation is actually the general form of the linear Doppler formula, and can be used to define Z in general. Applying (4.4) and (4.5) in the cosmological case now, we find the appealingly simple result that

$$1 + Z = \frac{S(t_r)}{S(t_e)} = \frac{S_r}{S_e}. \tag{4.6}$$

Although we derived the relation (4.6) by use of an approximation, we follow common practice by using it to define the cosmological redshift Z between any two points. Note that in the homogeneous isotropic cosmologies which we are studying, the redshift relation between any two points depends only on their separation in time, and is independent of their spatial relationship, as long as they are *causally connected* – close enough for light rays to have travelled between them within the age of the universe. Some thought on this point is rewarding, but the basis of the solution is that this is a general statement about the constancy of the speed of light.

There are many cosmological applications of the relation (4.6). For example, we can estimate the ages of observed structures if we know their redshifts. The most distant known quasars at the time of writing have $Z \simeq 5$, and, assuming that the universe has been matter dominated at least since the time when the distant quasars were formed, then, since $S \sim t^{2/3}$ for dust, $1 + Z = (t_r/t_e)^{2/3}$, so that for these objects $t_e \simeq 6^{-3/2} t_r$. Taking the present age of the universe estimated on the basis of the age profile of globular clusters (Peebles 1971), $t_{now} \sim 10^{10}$ yr, we find that the the most distant quasars were formed before the universe was less than one tenth of its present age, when it was only 10^9 years old. Of course, the observed object may no longer exist – the age we estimate is the age of the object when the light by which we see it left on its journey to our galaxy.

Observed redshifts are normally measured by obtaining the wavelength shift of known spectral lines in a distant object. In practice, this means taking a spectrum of the distant object and locating some identifiable spectral features. For relatively nearby galaxies, the well-known sodium pair can often be used, while for distant quasars the hydrogen-α lines are normally more visible. The fractional wavelength shift of the object's spectrum relative to a stationary reference spectrum is measured, and an equivalent redshift value Z is then calculated using (4.5) as a definition. However, they are often also described in terms of a *redshift velocity*, which is the recessional velocity whose linear Doppler effect z would give the same value, $z = Z$, as the measured spectral redshift. The confusing aspect of all this is that the redshift velocity can easily become greater than the speed of light. This practice can cause difficulty for the innocent reader, since the relationship between cosmic distance (as expressed by S) and cosmic velocity involves the Hubble parameter, which is not known with any high degree of accuracy at the present.

4.2 Cosmological Horizons

Fundamental to our understanding of the ways that physical laws operate in the universe is the concept of a *horizon*. Horizons in the cosmological sense derive their name by having the same operational properties as the horizons we see in our landscapes or seascapes: they delimit our zone of vision, and, by their existence, define what we can and cannot see as we look around us. Just as we can stand on a hill and have another hill hidden from our view beyond our horizon, so can we observe the universe from our galaxy and yet have other galaxies hidden from our view by one of our cosmological horizons. Such horizons share another property of horizons on earth, namely that of being dependent on the position of the observer. Let us now go on to demonstrate the existence of such horizons in our universe, and display some of their properties.

Since the speed of light is known to have, in accordance with all accepted physical theories, a fixed speed, $c =$ constant, and which may further be taken to be $c = 1$ by a choice of units, there is a maximum distance that light could have travelled in any finite time interval. In particular, since the universe is believed, on the basis of the best observational evidence, to have a finite age, this would be expected to have cosmological consequences. What is more, the speed of light is a limiting speed for all travellers and all signals. The practical consequence of this is that if two objects are too far away to have communicated by light signals (or any electromagnetic signal such as radio or microwaves) then they cannot have communicated by any other means. If two objects could not yet, in principle, have communicated with each other during the entire age of the universe, then they are called *causally disconnected*, on the grounds that if they could not have communicated in any way, then neither of them could possibly have caused any observable effect at the location of the other object.

To understand the physical consequences of these points, consider the path of a light ray, which is described by the constant-velocity law

$$\frac{\mathrm{d}x}{\mathrm{d}t} = c = 1 \iff \mathrm{d}x = \mathrm{d}t. \qquad (4.7)$$

Now in the comoving coordinates we have been using, the physical incre-

Cosmological Horizons

ment is $dx = S(t)dr$, where r is the (comoving) coordinate distance, and x is the physical distance. We can calculate the coordinate distance, r_p travelled by a light ray from the beginning of the universe up to a time t as

$$r_p = \int_0^t dr = \int_0^t \frac{dt'}{S(t')}, \tag{4.8}$$

which corresponds to a physical distance

$$d_p(t) = S(t)r = S(t) \int_0^t \frac{dt'}{S(t')}. \tag{4.9}$$

Pausing to examine this result, we see that d_p is the maximum straight-line distance that could have been travelled by a light ray since the beginning of the universe at $t = 0$. By the argument given above, this means that if we consider a particle with our coordinates centred on it, at the present, then $d_p(t)$ is the radius of a sphere enclosing all the other particles that could already have been seen – or could have been seen by – that particle. Because of this, d_p is known as the *particle horizon distance*. (See also Rindler 1956 for more details.)

For example, consider a model with no pressure and $\Omega = 1$. Then recalling from the solution to equation (3.22) that $S(t) \sim t^{2/3}$, one finds

$$d_p(t) = t^{2/3} \int_0^t \frac{dt'}{(t')^{2/3}} = 3t^{2/3}[t^{1/3}]_0^t = 3t. \tag{4.10}$$

The effect of the expansion of the universe is clearly contained in the factor of 3, since the result in a static universe would be expected to be simply $d = t$. One can think of the cosmic expansion as stretching out the physical distance already traversed by any light ray up to a given time. An equally valid way of interpreting this result is that the expansion carries the light rays along with it.

Notice also that it is perfectly possible that (4.9) results in a divergent expression for d_p. This means that no particle horizon exists anywhere in such a model, since any observer can already have seen past any sphere placed at a finite distance.

It is also possible to define another kind of horizon, defined in a similar way to (4.9). This is the *event horizon distance*

$$d_e = S(t) \int_t^\infty \frac{dt'}{S(t')}. \tag{4.11}$$

In contrast with the particle horizon, which encloses all the particles an observer *could have seen* at the present, the event horizon encloses all parts of the universe *which, in principle, can be reached* in the future by someone sending out signals at lightspeed or slower at a time t. Taking the same particular case for an example as before, with $\Omega = 1$ and $S \sim t^{2/3}$, we find

that
$$d_e(t) = t^{2/3} \int_t^\infty \frac{dt'}{(t')^{2/3}} = 3t^{2/3}[t'^{1/3}]_t^\infty \qquad (4.12)$$
which diverges, so there is no event horizon present in these models, and every observer can contact every other observer during the lifetime of the universe, provided they are sufficiently patient. This result could easily have been anticipated on the basis of the discussion we have already given, since the scale factor measures the physical distance between particles – and hence observers – and is growing like $t^{2/3}$ in a matter dominated universe, while the horizons all grow like t, so that the horizon is growing faster than the observers are expanding away from each other. Since the $\Omega = 1$ models expand forever, the event horizon will eventually encompass all points in these universe models.

4.3 Review

The redshifts of cosmological objects observed by Slipher and used by Hubble to build into his expansion law are to be understood as direct consequences of the way that the universe is expanding. The light of a photon is redshifted by the time it reaches a distant observer because the space through which it has travelled has been stretched out by the expansion of the universe.

Cosmic horizons exist for a similar reason to the cause of the cosmological redshift. If the expansion rate is very great, as it is on sufficiently large scales, then light may be dragged away from its intended destination faster than it can travel in that direction. Because this happens, there are parts of the universe which can never influence us and from which we can never receive any information – these lie beyond our particle horizon – and there are other parts to which we can never transmit any information or influence in any way – these lie beyond our event horizon.

4.4 Exercises

Exercise 4.1 *In our universe, galaxies expand like dust ($P = 0$). Assuming that the density parameter $\Omega = 1$, estimate how far away a galaxy would need to be in order to have a redshift velocity equal to the speed of light. How old would the universe have been when the spectral light left the galaxy?*

4.5 References

Peebles, P.J.E. *Physical Cosmology* (Princeton University Press, 1971).
Rindler, W. *Mon. Not. Roy. Astr. Soc.* **116**, 662 (1956).

5

Evolution of the Cosmological Density Parameter

This chapter is concerned with the question of whether our physical universe corresponds to one that will expand forever or one that will eventually halt its expansion and collapse back upon itself. Another important problem is to understand how close our universe is to the dividing case of $k = 0$, namely the model universe that only just continues to expand indefinitely.

5.1 The Flatness of the Universe

The results and discussion of Chapter 3 raise the obvious question of whether our own universe is gravitationally bound or unbound. If the former, the expansion will halt at some time in the future, and the universe will recollapse, while the latter possibility will result in permanent expansion. It turns out that this question is answerable in principle, but like so many other problems in the natural sciences, it is far simpler to provide the theoretical part of the solution than it is to discover which case applies to our own universe by direct observation. The answer will not be known with any confidence until dependable observational and theoretical results can be combined.

The conventional observational quantities used in the discussion of this section are, firstly, the *cosmological density parameter*

$$\Omega = \frac{8\pi G \rho}{3H^2}, \qquad (5.1)$$

which measures the ratio of potential to kinetic energy in the expansion: $\Omega = E_P/E_K$. Secondly, there is the *deceleration parameter*

$$q = -\frac{1}{H^2}\frac{\ddot{S}}{S}. \qquad (5.2)$$

The interpretation of these quantities is relatively straightforward – q measures the deviation of the growth of the cosmic scale factor from linearity as follows:

Take the Taylor expansion of S,

$$S(t) = S_0 + \dot{S}_0(t - t_0) + \frac{1}{2}\ddot{S}_0(t - t_0)^2 + \ldots \tag{5.3}$$

so that

$$\frac{S}{S_0} = 1 + H_0(t - t_0) - \frac{q_0}{2}H_0^2(t - t_0)^2 + \ldots \tag{5.4}$$

from which it is clear that q measures the deviation from linearity of the growth of the scale factor. The density parameter, as is clear from the derivation of the cosmological evolution equations given earlier, measures the ratio of potential to kinetic energy of the expansion. Another way of looking at this is to divide equation (3.19) by $3H^2$ to get

$$\Omega = 1 + \frac{K}{H^2}, \tag{5.5}$$

so that $\Omega > 1$ if the energy is negative, with $K > 0$, and $\Omega < 1$ if $K < 0$. The special case of zero total energy corresponds to $\Omega = 1$. Ω is a useful parameter because it is, in principle, straightforward to derive by measurement: one measures the present value, H_0, of the Hubble parameter, and the present mean mass density, ρ_0, to find Ω_0, the present value of the density parameter.* Also, the evolution of Ω with the expansion, or the time, is simply determined by a first-order differential equation, which will be obtained below. Lastly, the *horizon problem*, which is the question of why the universe appears the same everywhere we look in spite of the fact that looking in opposite directions one sees parts of it that appear never to have been in physical or causal contact (discussed elsewhere in this volume), can be formulated entirely in terms of Ω, without any explicit reference to time. Having emphasized the importance of examining the cosmological density parameter in understanding the universe, we turn now to the particular study of its evolution.

In order to progress further in studying the evolution of the universe, it is necessary to possess an equation of state. Using equation (3.10), we shall take this from now on to be of the *adiabatic form*

$$P = (\gamma - 1)\rho \tag{5.6}$$

which may be taken as the definition of a new quantity $\gamma(t)$, the *adiabatic index* of the cosmic fluid. This form is physically motivated by the fact that it describes accurately the two important cases of dust and radiation. Another reason for this choice is that, in many cases of physical interest, the adiabatic index is constant, making the analysis considerably simpler. Using the constant-γ form, we can show the relationship between q and Ω

---

* This is actually extremely difficult in practice, and there is little widespread agreement about which measurements and values are most plausible. An accessible discussion of the problems encountered in estimating Ω_{now} is given by Weinberg (1972).

The Flatness Approximation

by using equations (3.14) and (5.6) to write (Weinberg 1972)

$$\frac{\ddot{S}}{S} = \frac{1}{2}(2 - 3\gamma)\frac{8\pi G\rho}{3}, \qquad (5.7)$$

so by the definitions (5.1) and (5.2),

$$q = -\left(1 - \frac{3}{2}\gamma\right)\Omega. \qquad (5.8)$$

The case of pressureless matter, namely dust, corresponds to making the choice $\gamma = 1$, so that the deceleration parameter reduces to the simple form $q = \Omega/2$.

5.2 The Flatness Approximation

For the purpose of studying many cosmological questions, it is useful to assume that the universe is flat, with $\Omega = 1$ at all times. This simplifies many calculations because in this case the solutions to the expansion equations are radically simplified (see Chapter 3 above). However, one then needs to know how well, and during what epochs of cosmic evolution, this approximation holds. The answer to this question has a particularly neat formulation in terms of the density parameter and the cosmological redshift. We know that at early times the expansion equation (3.19) derived from energy balance will be dominated by the energy density of matter or radiation, since these scale as $\rho_m \sim S^{-3}$ and $\rho_r \sim S^{-4}$, so for small enough S these terms dominate the energy term $K \sim S^{-2}$. So the K term can be ignored until the time when its magnitude becomes comparable to the more rapidly decreasing matter energy density, $8\pi G\rho \sim 3K$. From equation (5.5) we have

$$K = \frac{k}{S^2} = (\Omega - 1)H^2 \qquad (5.9)$$

and in particular

$$K_0 = \frac{k}{S_0^2} = (\Omega_0 - 1)H_0^2 \qquad (5.10)$$

where H_0 is the present value of the Hubble parameter, and, likewise, for S_0, Ω_0, and K_0. Combining (5.9) and (5.10) gives

$$\frac{k}{S^2} = H_0^2(\Omega_0 - 1)\frac{S_0^2}{S^2}. \qquad (5.11)$$

Now since $\rho = \rho_0(S_0/S)^3$, the K and $8\pi G\rho$ terms become of comparable magnitude when

$$8\pi G\rho_0 \left(\frac{S_0}{S}\right)^3 \simeq \left\|3H_0^2(\Omega_0 - 1)\left(\frac{S_0}{S}\right)^2\right\|. \qquad (5.12)$$

Rewriting the scale factor parts of this relation in terms of the redshift Z, with $1 + Z = S_0/S$, transforms equation (5.12) to

$$\frac{8\pi G \rho_0}{3H_0^2} (1+Z)^3 \simeq \left\| (\Omega_0 - 1)(1+Z)^2 \right\|. \tag{5.13}$$

Since $8\pi G\rho/3H_0^2 = \Omega_0$, this says that the expansion of the universe is dominated by the energy density up to a redshift Z_f given by

$$1 + Z_f \simeq \left\| 1 - \frac{1}{\Omega_0} \right\|. \tag{5.14}$$

Up until Z_f, the K term can be ignored in the expansion equations. For example, if the present value of the density parameter is about 0.5, then $Z_f \simeq 0$, so that the expansion has been dominated by the energy density all the way up to the present. The result (5.14) will also be useful later on when we discuss the evolution of cosmic structure.

5.3 Relation of Density and Horizons

The density parameter also has an important role in understanding the question of the presence of horizons in the universe. There is an alternative form of the particle horizon distance which is independent of the time elapsed since the beginning of the universe, and depends only on knowing Ω as a function of the scale factor. From equation (4.9) we have

$$\begin{aligned} d_p(t) &= S(t) \int_0^t \frac{dt'}{S(t')} \\ &= S(t) \int_0^t \frac{dS}{S(t)\dot{S}(t)} \\ &= S \int_0^S dS\, S^{-2} (\Omega H^2 - K)^{-1/2} \ ; \ (\Omega \neq 1). \end{aligned}$$

The final result for the horizon distance as a function of the cosmic scale factor is

$$d_p(S) = S \int_0^S dS\, S^{-2} K^{-1/2}(S) \left[\Omega(S) - 1\right]^{1/2}. \tag{5.15}$$

This is useful because we know that $K \sim 1/S^2$, and the dependence $\Omega(S)$ is known in principle from either equation (5.23) or (5.27). When $\Omega = 1$, the original definition given in equation (4.9) has to be used directly, but this is the particular case where $S \sim t^{2/3\gamma}$ (for $K = 0$, $0 < \gamma \le 2$), and so the integral (4.9) can be performed directly.

5.4 The Evolution of the Density Parameter

Similar methods to those used in the previous section give an easy derivation of the evolution equation for Ω. The following derivation is based on that given by Madsen and Ellis (1988) (see also Madsen et al. 1992). Note that equation (5.6) along with (3.15) gives

$$\dot{\rho} = -3\gamma H \rho \qquad (5.16)$$

so that, in particular,

$$\gamma = \text{constant} \Rightarrow \frac{\rho}{\rho_0} = \left(\frac{S}{S_0}\right)^{-3\gamma} \qquad (5.17)$$

although in general there will be an *effective adiabatic index* $\gamma(s)$ which varies with the expansion. Now differentiating equation (5.1) with respect to time gives

$$\frac{\dot{\Omega}}{\Omega} = \left[\frac{\dot{\rho}}{\rho} - 2\frac{\dot{H}}{H}\right], \qquad (5.18)$$

and substitution from equation (5.16) yields

$$\dot{\Omega} = -\Omega H \left[3\gamma + 2\frac{\dot{H}}{H^2}\right]. \qquad (5.19)$$

Now equation (3.14) can be written as

$$3\dot{H} + 3H^2 + \frac{1}{2}(3\gamma - 2)8\pi G\rho, \qquad (5.20)$$

so that

$$\frac{\dot{H}}{H^2} = -1 - \frac{1}{2}(3\gamma - 2)\Omega, \qquad (5.21)$$

and we finally obtain

$$\dot{\Omega} = (2 - 3\gamma)H\Omega(1 - \Omega). \qquad (5.22)$$

This equation is the fundamental evolution equation, and is conceptually equivalent to equation (3.14). It is far easier to study the behaviour of Ω, though, if we introduce a new parametrization of the equations by defining the *logarithmic scale factor* $s = \ln S$, so that $\dot{s} = H$, and we can thus deduce

$$\frac{d\Omega}{ds} = (2 - 3\gamma)\Omega(1 - \Omega). \qquad (5.23)$$

Equation (5.23) is very useful to us, because we can avoid the necessity of knowing the exact dependence of any of the dynamical quantities on time. Furthermore, it is linear in the derivatives of Ω, because the nonlinearity of equation (3.13) is masked by the definition (5.1) of Ω. Use of equation (5.23) allows us to plot curves showing the behaviour of $\Omega(s)$, provided only that $\gamma(s)$ is specified. We shall do this in a moment, but first notice that

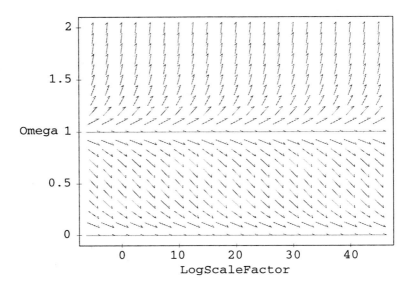

Figure 5.1 *Phase diagram showing how $\Omega(s)$ varies with the logarithm of the cosmic scale factor when $\gamma > 2/3$.*

we can easily write down an exact solution for $\Omega(s)$ when $\gamma = constant$. From equations (3.19) and (5.1), one has

$$\Omega = \frac{8\pi G\rho}{8\pi G\rho - 3K} \Rightarrow \quad (5.24)$$

$$\Omega = \left[1 - \frac{3K}{8\pi G\rho}\right]^{-1}, \quad (5.25)$$

and for $\gamma = constant$, we see

$$\frac{3K}{8\pi G\rho} = \frac{3K}{8\pi G\rho_0}\left(\frac{S_0}{S}\right)^2\left(\frac{S}{S_0}\right)^{3\gamma} = \left(\frac{3K_0}{3H_0^2}\right)\left(\frac{3H_0^2}{8\pi G\rho_0}\right)\left(\frac{S_0}{S}\right)^{2-3\gamma}. \quad (5.26)$$

Equation (5.5) shows that $3K/3H^2 = \Omega - 1$, so putting it all together we find the exact solution

$$\Omega(S) = \left[1 + \left(\frac{1 - \Omega_0}{\Omega_0}\right)\left(\frac{S_0}{S}\right)^{2-3\gamma}\right]^{-1}. \quad (5.27)$$

Even though this solution only holds when γ is constant, equation (5.27) is of little use, since it is difficult to draw its graph. For the purpose of our analysis equation (5.23) has the dual advantages of being more general, and allowing a simpler visualization of the behaviour of Ω with the cosmic expansion.

From equation (5.23) we see that $\Omega = 1$ is a constant solution for all γ, and that for $\gamma = 2/3$, all values $\Omega = \Omega_0 = constant$ are always solutions. Next we notice that $d\Omega/ds > 0$ if $\gamma > 2/3$ and $\Omega > 1$, or if $\gamma < 2/3$ and $\Omega < 1$, and that $d\Omega/ds < 0$ if $\gamma < 2/3$ and $\Omega > 1$, or if $\gamma > 2/3$ and $\Omega < 1$. We can summarize these statements by saying that when $\gamma > 2/3$, Ω diverges away from unity, and Ω approaches unity for $\gamma < 2/3$. Since $\Omega = 1$ is a *fixed point* of the solution, this shows that $\Omega = 1$ is also an *asymptote* of all the solutions for $\gamma < 2/3$, and an *unstable past asymptote* for $\gamma > 2/3$. In particular, for pressureless matter, with $\gamma = 1$, the value of Ω diverges from 1 quite rapidly. Also for all $\gamma > 2/3$, if $\Omega > 1$, then $d\Omega/ds \to \infty$ for finite s – clearly this happens at the point when these models stop expanding and turn around, since there $\rho > 0$ and $H = 0$, so $\Omega = \infty$. Of course, $\Omega = 0$ is another constant solution to (5.23), but this means that $\rho = 0$, and we are less interested in this possibility, since it corresponds to a model of a universe empty of matter. The physical content of the preceding discussion is encapsulated by Figures 5.1 and 5.2. The first of these two graphs shows that if, as we believe, the universe can be well modelled as filled with homogeneous pressureless matter having adiabatic index $\gamma = 1$, then it would be surprising if Ω was close to unity now, since examination of the curves would lead us to expect that the expansion of the universe has caused the value of Ω to diverge a long way from this value by now. The fact that the observational evidence restricts the density parameter to lie in the range $1/10 \leq \Omega \leq 10$ therefore gives rise to the *flatness problem*, as it has come to be known, since it is equivalent to the observation that the *Gaussian curvature measure*, k/S^2, of space is remarkably close to zero. This problem, and its solution, will be discussed in greater detail in the section preceding our treatment of cosmological inflation. The very existence of the flatness problem, however, shows that it is very difficult to decide whether the universe is bound or unbound. The future evolution of the universe is a question which therefore poses severe physical problems as well as metaphysical ones.

5.5 Review

The cosmological density parameter provides a convenient way of looking at the evolution of the mass density of the universe, and generates a clear statement of the cosmological flatness problem. The early stages of the expansion of the universe are like those of a flat model, and this approximation holds until the redshift is approximately the inverse of the density parameter. A differential equation for the evolution of the density

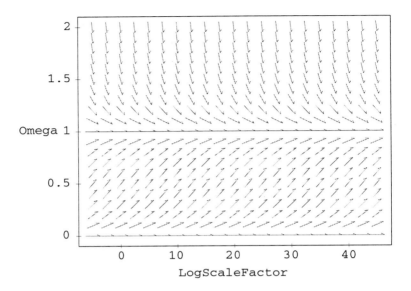

Figure 5.2 *Phase diagram showing how $\Omega(s)$ varies with the logarithm of the cosmic scale factor when $\gamma < 2/3$.*

parameter with the expansion is easily derived. This equation yields information about the way that the density parameter behaves more simply than the exact form of the density parameter. The horizon distance can also be simply formulated in terms of the density parameter, allowing the expression of the horizon distance in terms of the cosmic expansion factor. As for the density parameter, this means that the existence of horizons can be investigated without the need to construct a solution for the expansion of the universe with time.

5.6 Exercises

Exercise 5.1 *Verify that equation (5.27) is the solution to (5.23).*

Exercise 5.2 *Convince yourself that the equation for $d\Omega/dS$ holds regardless of whether the adiabatic constant γ of the matter varies or not. Now assume that γ remains constant as S varies. Then by considering the second derivative of Ω, categorize the curves in the (S, Ω) phase plane. When is $d\Omega/dS$ a maximum or minimum, and for what range of values of γ does this occur? Extend the phase plane diagrams shown in Figures 5.1 and 5.2* to include collapsing universe models. Consider a collapsing universe as being an expanding one running backwards.

Exercise 5.3 *Classify all expanding universe models filled with matter whose equation of state is of the form $P = (\gamma - 1)\rho$, where $0 \leq \gamma \leq 2$ and values of Ω are arbitrary, according to whether they have a*

1. *particle horizon;*
2. *event horizon.*

5.7 References

Madsen, M.S. and Ellis, G.F.R. *Mon. Not. Roy. Astron. Soc.* **234**, 67 (1988).

Madsen, M.S., Mimoso, J.P., Butcher, J.A. and Ellis, G.F.R. *Phys. Rev. D* **46**, 1399 (1992).

Weinberg, S. *Gravitation and Cosmology* (Wiley, New York, 1972).

6

The Thermal History of the Universe

In this chapter, we will begin to use the knowledge gained in earlier sections to construct models of the physical processes which have taken place during the history of our universe. Essential to furthering our understanding is the *thermal history* of the universe, by which we mean the relationships between the age and size of the universe, and its average temperature. It turns out that we can make a lot of progress towards understanding this thermal history on the basis of only a few well-based assumptions.

6.1 The State of Matter in the Early Universe

In an earlier section, in equation (5.6), we introduced the adiabatic index γ as a convenient way of parametrizing an equation of state with a linear relationship between P and ρ, $P = (\gamma - 1)\rho$. This equation of state is of interest because it includes the two most important cases for study of the evolution of the universe, namely that of *pressureless matter*, or *dust*, with $P = 0 \Leftrightarrow \gamma = 1$, which we have already encountered in earlier sections, and *radiation*, which has $P = \rho/3 \Leftrightarrow \gamma = 4/3$. The equation of state (5.6) is also useful, because it allows us to determine $\rho(S)$ for any value of γ by solving equation (3.15), since with application of (5.6) this becomes

$$\frac{\dot\rho}{\rho} = -3\gamma\frac{\dot S}{S}. \tag{6.1}$$

In the case that γ is constant, this last equation is directly integrated to obtain

$$\rho(S) = \rho_0 \left(\frac{S}{S_0}\right)^{-3\gamma}. \tag{6.2}$$

Equation (6.2) shows explicitly that γ is an *adiabatic* index, since any typical volume element of the universe scales as $V \sim S^3$, and hence $\rho \sim V^{-\gamma}$, that is, the energy density of the cosmic fluid scales as a power of its volume. This is essentially in accordance with the meaning of the term 'adiabatic', because it shows that the fluid conserves the total energy contained within any given expanding element of volume.

In the particular case of electromagnetic radiation, $P = \rho/3$, so $\rho \sim S^{-4}$, or equivalently $\rho \sim V^{-4/3}$. Electromagnetic radiation in thermal equilibrium at a temperature T has an energy density $\rho \sim T^4$, so we can see that if the expansion is adiabatic, then the radiation temperature $T \sim 1/S$, or

$$T(S) = \frac{T_0 S_0}{S}. \tag{6.3}$$

The fact that the radiation energy density falls off more rapidly than the matter energy density means that if we follow the expansion backwards in time, the radiation energy density comes to dominate completely, since $\rho_m \sim 1/S^3$, $\rho_r \sim 1/S^4$, and hence $\rho_m/\rho_r \to 0$ as $S \to 0$. Also, since $\rho_r \to \infty$ as $S \to 0$, the model cosmos we have developed so far is seen to start with a *hot big bang*.

It is also very important to notice that both the matter and radiation energy terms dominate the 'binding energy' term – namely the $-3k/S^2$ appearing on the right of equation (3.13) – at early times when S is small, so that it can be neglected in discussions of the earliest stages of the cosmological expansion. The reasoning is clearly along the same lines as used above to compare the matter and radiation energies. In later sections, however, it will be useful to realize that the $-3k/S^2$ term acts exactly as if it were the contribution to the total energy density from a fluid with pressure $P_k = -\rho_k/3 \Leftrightarrow \gamma_k = 2/3$, with the one peculiarity that ρ_k may be either positive or negative. There is no contribution from this fluid to equation (3.14) since $\rho_k + 3P_k = 0$, and (6.2) yields the prediction that $\rho_k \sim 1/S^2$, so that we can verify the consistency of this viewpoint.

6.2 The Cosmic Entropy Density

Deducing the thermal history of the universe depends largely on knowing how the entropy behaves. Intuitively, the entropy within each comoving volume should be conserved, since there is no outside source of energy to heat any region of the universe as it expands. However, the justification of this statement can be made on solid thermodynamic grounds. In particular, we are able to calculate the exact form of the entropy function for a comoving volume. Since the calculation is instructive, we shall present it before proceeding to deducing the thermal aspects of cosmic evolution.

Consider the first law of thermodynamics relating the energy E, temperature T, entropy Σ and pressure P of a gas contained in a volume V:

$$dE = Td\Sigma - PdV. \tag{6.4}$$

The Cosmic Entropy Density

By definition $E = \rho V$, so this can be rewritten as

$$d\Sigma = \frac{1}{T}[d(\rho V) + P dV]. \tag{6.5}$$

Now setting $d\rho = 0$ in equation (6.5) and dividing by dV one has

$$\frac{\partial \Sigma}{\partial V} = \frac{1}{T}(\rho + P) \tag{6.6}$$

while setting $dV = 0$ in (6.5) and dividing by dT results in

$$\frac{\partial \Sigma}{\partial T} = \frac{V}{T}\frac{d\rho}{dT}. \tag{6.7}$$

Now the integrability condition for these equations is that the partial derivatives should be independent of order, so that

$$\frac{\partial^2 \Sigma}{\partial T \partial V} = \frac{\partial^2 \Sigma}{\partial V \partial T}. \tag{6.8}$$

Taking the appropriate partial derivatives of equations (6.6) and (6.7), the integrability condition (6.8) gives the dependence of the pressure on temperature:

$$\frac{dP}{dT} = \frac{1}{T}(\rho + P). \tag{6.9}$$

Now the law of conservation of energy at constant entropy is, from equation (6.4)

$$d(\rho V) + P dV = 0, \tag{6.10}$$

from which we deduce the way that the pressure varies with time:

$$\frac{dP}{dt} = \frac{1}{V}\frac{d}{dt}[(\rho + P)V]. \tag{6.11}$$

Clearly, we can combine equations (6.9) and (6.11) to get

$$\frac{dP}{dt} = \frac{dP}{dT}\frac{dT}{dt} \implies \frac{dP}{dt} = \frac{1}{T}(\rho + P)\frac{dT}{dt}. \tag{6.12}$$

Now substituting from equation (6.10) gives

$$\frac{V}{T}(\rho + P)\frac{dT}{dt} = \frac{d}{dt}[(\rho + P)V]. \tag{6.13}$$

Equation (6.13) has an integrating factor of $1/T$, allowing it to be written as

$$\frac{d}{dt}\left[\frac{V}{T}(\rho + P)\right] = 0. \tag{6.14}$$

The next step is to notice that equation (6.5) can be rewritten in the form

$$d\Sigma = \frac{1}{T}d[(\rho + P)V] - \frac{V}{T}dP, \tag{6.15}$$

and that substituting in this for dP from equation (6.9) gives

$$d\Sigma = \frac{1}{T} d\left[(\rho + P)V\right] - V(\rho + P)\frac{dT}{T^2} \tag{6.16}$$

which can be directly integrated to give

$$d\Sigma = d\left[\frac{V}{T}(\rho + P)\right]. \tag{6.17}$$

This means that, up to an arbitrary additive constant, the entropy in a comoving volume V is

$$\Sigma = \frac{V}{T}(\rho + P). \tag{6.18}$$

Furthermore, from equation (6.14) it is obvious that this entropy is constant throughout the expansion,

$$\frac{d\Sigma}{dt} = 0. \tag{6.19}$$

Note that although this last conclusion was actually built into the calculation through the form (6.10), the result of the calculation is the general form (6.18) for the entropy function in an expanding universe.

6.3 How the Universe Cooled

The detection of the cosmic microwave background radiation by Penzias and Wilson (1965) and the theoretical explanation of its origin by Dicke et al. (1965) shows that there is present in the universe a small energy density contribution from thermal blackbody radiation. This section will present and interpret these results in the light of the expanding universe model developed in earlier sections.

It was demonstrated in section 6.1 that the universe began by expanding from a singularity where the energy density was infinite. The argument that ρ_r dominated ρ_m at early times, shows that immediately after the singularity, the radiation energy density was dominant, with a large temperature which also becomes infinite at the singularity. This singularity was therefore a *hot big bang*. Since we are unable, in either principle or fact, to reconstruct any events prior to the big bang, we are justified in choosing the time of the bang to be the origin of our time, $t = 0$, and, therefore, in measuring the age of the universe from this instant.

The evolution of the radiation energy density and temperature will be known from equations (6.2) and (6.3) if we can solve for $S(t)$. The energy density of thermal radiation is known to be proportional to the fourth power of the temperature,

$$\rho_r = aT^4, \tag{6.20}$$

with the constant a being known as the *blackbody constant*. Furthermore, in natural units, which are chosen so as to have $k = \hbar = c = 1$, the value

How the Universe Cooled

of a simplifies to $a = \pi^2/15$ (see Appendix A). At this early time, since $\rho_r \gg \{\rho_m, \rho_k\}$, (3.13) can be written

$$\dot{S}^2 = \frac{8\pi Ga}{3} S^2 T^4 = \frac{8\pi^3 G}{45} S^2 T^4. \tag{6.21}$$

By use of equation (6.2) for ρ_r, this shows that

$$S\dot{S} = q = \text{constant}, \tag{6.22}$$

which has the solution

$$\frac{1}{2} S^2(t) = q(t - t_0) \Rightarrow S(t) = \sqrt{2q}\, t^{1/2} \tag{6.23}$$

where q is a constant and we have chosen $t_0 = 0$ so that $S(t_0) = 0$ as agreed above. Further progress can be made by noticing that if the Boltzmann constant, k, is chosen to be unity, then energy and temperature are equivalent and, measuring temperature in MeV and time in seconds, we find that

$$t \simeq 3 \left[\frac{1\text{ MeV}}{T}\right]^2 \text{ sec}, \tag{6.24}$$

so that by the time the universe is just one second old, it has cooled to a temperature of order 1 MeV. Since 1 MeV is a typical energy scale for nuclear reactions, we might expect that nuclear reaction rates would be important for the evolution of the universe at a time of order a few seconds after the bang. As we shall discover below, these expectations are entirely justified.

Using the rough estimate given earlier that the age of the universe is about $t_{now} \sim 10^{10}$ yr $\sim 3 \times 10^{17}$ sec, equation (6.24) gives $T_{now} \sim 10^{-9}$ MeV ~ 1 K, which is in fair agreement with the measured value $T_{now} \simeq 3$ K. The hot big bang thus allows us to understand the presence of a 3 K bath of thermal radiation in the universe today. This radiation is the *relic* left after the universe has expanded, allowing the hot photon gas comprising the radiation to cool from the extreme temperature of the big bang down to the cold present.

From this discussion, we know that the energy density of the universe was dominated by thermal radiation at and after the big bang, but we also know that it is presently the matter that dominates, since a temperature of $T \sim 3$ K corresponds to $\rho_r(now) \sim 10^{-34}$ g/cm^3, while $\rho_m(now) \sim 10^{-29}$ g/cm^3, so that at the present

$$\left.\frac{\rho_m}{\rho_r}\right|_{now} \sim 10^5. \tag{6.25}$$

When did the transition take place? Recall that from the discussion following equation (6.3), the redshift is given by $\rho_m/\rho_r \sim S \sim 1 + Z$, so (6.25) tells us that ρ_m and ρ_r were of comparable magnitudes when S was about

$10^{-5} S_{now}$, or when the universe was about one hundred thousand times smaller than it is now. If the matter-dominated solution with $1 + Z \sim t^{2/3}$ has been valid since then, we have that

$$10^5 \sim \frac{S_{now}}{S_{then}} \sim (1+Z) \sim \left(\frac{t_{now}}{t_{then}}\right)^{2/3}, \tag{6.26}$$

or $t_{now} \sim 3 \times 10^7 t_{then}$, so that the age of the universe when the matter dominated era began was about $t_{then} \sim 100$ to 1000 yr.

6.4 The Neutrino Temperature

Interestingly, since all the particles that make up the universe were once in thermal equilibrium, one would initially expect that all massless particles would always have the same temperature. However, this expectation turns out to be mistaken, although only in a small way. Nevertheless, the physical process by which there come to be temperatures different from that of the radiation background is sufficiently interesting that we shall examine it here as an example of the kind of physics which makes cosmology subtle and fascinating.

The physical reason for there being two temperatures at the present is that the neutrinos, although in thermal equilibrium with the radiation at high temperatures, decouple from the other contents of the universe before the time at which the equilibrium numbers of electrons and positrons undergo mutual annihilation. Neutrino decoupling happens because the weak interaction rates scale steeply with temperature – the cross-sectional interaction rate per particle $\langle v\sigma \rangle \sim G_F^2 T^2$, where G_F^2 is the Fermi weak interaction constant, v is the average particle velocity, and σ is the mean collision cross-section. Since the number density of neutrinos $n_\nu \sim T^3$ as for other species, the total weak interaction rate is like $G_F^2 T^5$. In order to see this another way, consider what happens when weak interactions take place:

The strength of the collision is measured by the Fermi factor G_F^2, where G_F is the Fermi constant, and there is one factor of G_F for each particle. Also, these interactions need to conserve energy. The overall reaction rate is therefore constrained by a Dirac delta function to involve only those particles that are in areas of phase space permitted by energy conservation. The relevant phase space volume is like $V \sim k^3$, where k is the particle momentum, so that the phase space volume element per particle is like $dV \sim k^2 \, dk$. Putting this together, the reaction rates, λ_{ij}, between particle species i and j are all of similar form to

$$\lambda_{ij} \sim \int dk_i \int dk_j \, \delta(k_i - k_j) k_i^2 G_F^2 k_j^2 \tag{6.27}$$

The Neutrino Temperature

$$\sim \int dk\, G_F^2 k^4 \qquad (6.28)$$

$$\sim k^5 G_F^2 \qquad (6.29)$$

$$\sim G_F^2 T^5. \qquad (6.30)$$

The last line follows from the fact that the particles all have thermal energies as mentioned above. So for temperatures greater than the particle mass, $T \geq m$, the reaction rate coefficients λ are high enough to maintain those particles in equilibrium, since the expansion timescale is $1/H \sim t \sim T^{-2}$ and the corresponding rate is like $1/t \sim T^2$, which is far slower than the weak interaction rate.

These are the only interactions that couple the neutrinos to the other particles, so the neutrinos will decouple from the radiation when the weak interaction rate becomes slower than the expansion rate, at a freeze-out temperature, T_f, given by

$$G_F^2 T_f^5 \sim H \sim \frac{T^2}{M_P} \implies T_f \sim \left(\frac{M_P}{G_F^2}\right)^{1/3}. \qquad (6.31)$$

Notice that this result is independent of the exact properties of the neutrinos themselves: it depends only on the strength of the weak interactions as measured by the Fermi constant. This is because these are the only interactions through which the neutrinos can interact with other particles. Above T_f, the neutrinos remain in equilibrium with everything else, but below T_f, the weak interactions are too slow to maintain equilibrium with the radiation. The result of this is that the neutrinos carry on expanding freely, their temperature still scaling in the same way as that of the radiation, $T_\nu \sim T_{rad} \sim 1/S$, so that although they are not in contact with the radiation, their temperature is tracking that of the other species.

Now comes the twist. At temperatures just below about 1 MeV, there is no longer sufficient thermal energy in collisions to create electron–positron pairs, since the electron mass $m_e \simeq 0.5$ MeV. The existing electrons and positrons present annihilate each other, heating the radiation. This happens because the radiation is receiving energy from the electron–positron gas, but cannot give any back through the creation of electron–positron pairs when $T < m_e$. The amount of heating of the radiation can be estimated from the fact that the entropy given by equation (6.18) remains constant throughout the annihilation phase.

The number of spin degrees of freedom in the electron–positron gas is four (two spin states for electrons and two for positrons), while that for the photons is two. So the contributions of the electrons, e, and the positrons, $\bar{e}$, are each equal to that of the photons:

$$\rho_e = \rho_{\bar{e}} = \rho_\gamma \;;\; P_e = P_{\bar{e}} = P_\gamma = \frac{1}{3}\rho_\gamma. \qquad (6.32)$$

Using the expression (6.18) for the entropy before annihilation and substituting $\rho_\gamma = aT^4$, one has

$$\Sigma_{before} = \frac{V}{T}(\rho_\gamma + \rho_e + \rho_{\bar{e}} + P_\gamma + P_e + P_{\bar{e}}) = 4a(ST)^3 \tag{6.33}$$

whereas after annihilation the corresponding expression is

$$\Sigma_{after} = \frac{V}{T}(\rho_\gamma + P_\gamma) = \frac{4a}{3}(ST)^3. \tag{6.34}$$

Since $\Sigma_{before} = \Sigma_{after}$, the temperature of the radiation must increase due to the heating effect of the electron–positron annihilation in such a way that

$$\frac{T_{after}}{T_{before}} = 3^{1/3}. \tag{6.35}$$

Since $T_\nu = T_{before}$, this relation allows us to deduce that since the time of electron–positron annihilation, the radiation and neutrino temperatures have been related by

$$T_\nu = \frac{1}{\sqrt[3]{3}} T_{rad}. \tag{6.36}$$

There is therefore expected to be a cosmic neutrino background consisting of thermal neutrinos (neutrinos in thermal equilibrium with each other) at a present day temperature

$$T_\nu|_{now} \simeq 2\,\text{K}. \tag{6.37}$$

Of course, this neutrino sea which pervades the entire universe is still entirely undetectable by direct observations, due to the weakness of neutrino interactions with other particles. However, if these neutrinos possess even the tiniest mass, then their large number density, $n_\nu \sim n_\gamma$, ensures that they will have important consequences for the evolution of structure on large scales. This topic is discussed further in Chapter 11.

6.5 Decoupling and Recombination

At high temperatures, the universe was hot enough that the matter was completely ionized: the photons in the radiation background had enough energy on average to keep the hydrogen ionized. In particular, this means that the universe was opaque during this early phase, because there would have been a density of free electrons high enough that the photon mean free path, λ_e, would have been very short: $\lambda_e \simeq 1/n_e \sigma_e$, where σ_e is the Thomson scattering cross section of the electron, and $n_e \sim n_b$. As the universe cooled, however, the photon average energy T eventually dropped below that required to ionize a hydrogen atom, so that the electron–proton plasma could have recombined into atomic hydrogen. The matter and radiation had then decoupled from each other. At this stage, with no free electrons to scatter the photons, the photon mean free path became unbounded, so that the universe has been transparent since that epoch.

The simplest way to estimate the epoch of decoupling is to say that it took place at a temperature $T_{dec} = 13.6\,\text{eV}$, corresponding to the ionization energy of hydrogen. In this approximation, decoupling was a single event which occurred instantaneously at a particular time. For many purposes this approximation is sufficient. A more subtle approach, which is nearly as simple, is to realize that decoupling took place gradually over the range of temperatures corresponding to the hydrogen ionization energy. To estimate this range, notice that at any temperature the fractional number density of photons in the radiation background with sufficient energy to ionize hydrogen is given by the Boltzmann factor $\exp(-I/T)$, where $I = 13.6\,\text{eV}$ is the ionization energy of hydrogen:

$$\frac{n_i}{n_\gamma} = \exp\left(\frac{I}{T}\right). \tag{6.38}$$

Here n_i is the number density of ionizing photons. This can be rewritten as

$$\frac{n_i}{n_e} = \frac{n_\gamma}{n_b}\exp\left(\frac{I}{T}\right), \tag{6.39}$$

since by this time $n_e = n_b$, because $T_{dec} \ll m_e$ so that there is no excess of electrons beyond the number corresponding to the baryons present. The hydrogen will remain totally ionized as long as there is about one ionizing photon per electron, while it will begin to be only partially ionized as n_i/n_e drops below unity. From equation (6.39), this happens at a temperature

$$T_{dec} = \frac{I}{\ln(n_\gamma/n_b)}. \tag{6.40}$$

This calculation shows that the naive decoupling temperature, T_{dec}, is the temperature at which decoupling *begins*: the total separation of matter and radiation is only complete when the temperature is sufficiently lower than T_{dec}, so that $n_i/n_e \ll 1$.

6.6 Review

We can now construct a rough guide to the thermal history of the universe. At the earliest times, immediately after the big bang, the universe was filled with a dense hot radiation gas which emerged directly from the big bang itself. Other particles were also in thermal equilibrium with the radiation, and behaved as additional radiation gases, each with the same temperature as that of the radiation. After the weak reactions froze out, the neutrinos decoupled from the other particles and continued to expand freely. When the temperature fell below the electron mass, the electron–positron gas fell out of thermal equilibrium, and the electrons annihilated with the positrons, heating the radiation and raising its temperature above that of the neutrino gas.

As the universe expands, the radiation cools adiabatically, but continues to dominate the energy density of the expansion for about the first thousand years. After this time, the radiation era ends and the matter eventually comes to dominate the expansion: from this time on, the universe is in the matter era. Some time after the matter comes to dominate, the radiation decouples from the matter and the universe becomes transparent for the first time.

6.7 Exercises

Exercise 6.1 *The ionization energy of the hydrogen atom is $13.6\,eV$. Estimate the redshift at which the protons and electrons would have formed into atomic hydrogen. What was the age of the universe when recombination occurred? Can you think of any possible observable consequences of this transition?*

Exercise 6.2 *The discussion preceding equation (6.25) assumed that $\Omega_0 = 1$. Generalize it to allow for arbitrary values of Ω_0 and thus find the relation between the redshift, Z_m, at which the matter dominated era began and the present value of the density parameter Ω_0. Hint: refer to equation (5.27) to obtain the dependence of Ω on redshift.*

Exercise 6.3 *The calculation of the present day neutrino temperature given in equations (6.32–6.37) is a slightly simplified approximation. In fact, an electron or positron in a thermal electron–positron gas has its effective number of spin states slightly suppressed compared to a free electron or positron. This is because of the Pauli exclusion principle operating on each electron in the presence of a large number of other electrons. It turns out (for example, see Weinberg 1972) that each electron or positron in the particle gas has only 7/4 effective degrees of freedom. Improve the calculation of equations (6.32–6.37) to take account of this modification, and explain why the approximation used in the text is nearly as good.*

6.8 References

Dicke, R.H., Peebles, P.J.E., Roll, P.G. and Wilkinson, D.T. *Astrophys. J.* **142**, 414 (1965).

Penzias, A.A. and Wilson, R.W. *Astrophys. J.* **142**, 419 (1965).

7

Cosmological Synthesis of Elements

The most influential prediction of the expanding hot universe was that of the cosmological mass fraction of helium. It was known that the helium fraction was too large to be explained by production in stars within the age of the universe. The calculation of the production of the helium mass fraction in a hot early stage of the universe was stimulated by the discovery of the cosmic background radiation, and led to widespread acceptance of the hot universe model. The calculation is particularly interesting because it involves a variety of physical effects and processes. The calculation presented here also attempts to show how the result synthesizes and depends upon a variety of research.

7.1 Time of Element Synthesis

Let us begin this section by recalling that, according to equation (6.24), the age of the universe in its earliest stages is given by $t \sim (1\,\text{MeV}/T)^2$ sec, and that the energy density is dominated by the hot, dense, thermal radiation at temperature T. Since most nuclei have binding energies of at most a few MeV, we would not expect the universe to have contained any elements during the time $t \leq$ a few sec. Presumably there would instead have been only free neutrons and protons, and possibly a few other heavy particles, such as pions, in addition to the radiation. There would also have been large numbers of those particle species which were effectively massless at that stage, namely those with masses $\leq T$. For example, the electron mass $m_e \simeq 1/2$ MeV, so we would expect electrons and positrons to be present in the same numbers as the thermal photons, and also with a thermal spectrum – since they would be in thermal equilibrium with the radiation. It turns out that the exact numerical results are sensitive to the precise number of species present, but that the qualitative picture is remarkably *insensitive* to these considerations, so that we are able to paint an accurate picture of the element formation process without going into so much detail as to obscure the generality of the physics.

7.2 The Neutron to Proton Ratio

In order to understand what happens during the nucleosynthesis era, we need to understand the initial conditions for that era: what does the universe look like just before the elements are made? In particular, the final abundances of the elements heavier than hydrogen will obviously depend on the number of neutrons present, since there is no way to construct stable forms of such elements in the absence of neutrons. Thus, we are immediately led to examine the ratio of neutron to proton number as a function of the cosmic temperature. An added advantage of this way of approaching the problem, as it will turn out, is that it allows us to disregard complications which only alter the rate of cosmic expansion since, in the absence of such complications, we can always use equation (6.24) to recover the time dependence of the entire process.

We now introduce some notation: let X_n and X_p denote the fractional number densities of *neutrons* and *protons* with respect to the total number of *baryons*:

$$X_n = \frac{n_n}{n_B}; \ X_p = \frac{n_p}{n_B}; \ n_B = n_n + n_p. \tag{7.1}$$

Note that although, as has been remarked, there may be other particles present, the only stable baryons are the protons and neutrons. Antiprotons and antineutrons are present only in negligible numbers at the temperatures we are considering in this chapter – Chapter 8 discusses where the baryons themselves originally came from.

Without any detailed knowledge of the physics involved, we can construct an equation for X_n:

$$\frac{dX_n}{dt} = \lambda_{pn}(1 - X_n) - \lambda_{np}X_n \tag{7.2}$$

where we have used $X_p = (1 - X_n)$ from equation (7.1), and $\lambda_{np}, \lambda_{pn}$ are respectively the rates of conversion from neutrons to protons and vice versa. Clearly, we could write down a similar equation for X_p, but since $X_n + X_p = 1$, the solution to (7.2) determines both of the fractional number densities. The way we have written equation (7.2) ensures that all the complicated physical details are contained in the forms of the functions describing the conversion rates λ_{np} and λ_{pn}, so that a full solution to (7.2) requires knowledge of these rates as a function of temperature. Even without this knowledge, however, we can find the equilibrium solution to equation (7.2), defined as that with $dX_n/dt = 0$, which is

$$X_n = \frac{\lambda_{pn}}{\lambda_{pn} + \lambda_{np}}. \tag{7.3}$$

The Neutron to Proton Ratio

Now it can be demonstrated that at high temperatures, $T \geq m_B$, the weak interaction rate $\lambda \sim T^5$. The argument is the same as that applied to neutrinos in Chapter 6, but is applied to neutrons and protons here. Consider the volume of phase space available to a thermal particle of mass m and momentum k with $k \gg m$, in particular $k \sim T$. This phase volume is $\sim k^3$, so the number of particles with momentum k is like k^3. In particular, the number densities of protons and neutrons are found from

$$dN_n \sim k_n^2 dk_n$$

and

$$dN_p \sim k_p^2 dk_p.$$

When interactions between neutrons and protons take place, the strength of the collision is measured by the Fermi factor G_F^2 and, again, there is one factor of G_F for each particle. Also, these interactions need to conserve energy. The overall reaction rate is therefore constrained by a Dirac delta function to involve only those particles that are in areas of phase space permitted by energy conservation. Putting this together, the reaction rates are all of a form similar to

$$\lambda_{np} \sim \int dk_n \int dk_p \delta(k_n - k_p) k_n^2 G_F^2 k_p^2 \tag{7.4}$$

$$\sim \int dk\, G_F^2 k^4 \tag{7.5}$$

$$\sim k^5 G_F^2 \tag{7.6}$$

$$\sim G_F^2 T^5. \tag{7.7}$$

The last line follows from the fact that the particles all have thermal energies as mentioned above. So for temperatures greater than the neutron–proton mass, $T \geq 1\,\text{GeV}$ ($\simeq 10^{13}\,\text{K}$), the reaction rate coefficients λ are high enough to maintain equilibrium between neutron and protons, since the expansion timescale is $1/H \sim t \sim T^{-2}$ and the corresponding rate is like $1/t \sim T^2$, which is far slower than the weak interaction rate. The remarkable result, found by Hayashi (1950), is that at these temperatures, $\lambda_{pn} = \lambda_{np}$, so that in equilibrium

$$X_n = \frac{1}{2} = X_p. \tag{7.8}$$

Equation (7.8) provides a simple, and useful, initial condition for our study of nucleosynthesis. We need not worry about the precise evolution of the cosmos before the nucleosynthesis era: it suffices to know that, in accordance with our model of a hot big bang, the universe was hot enough that neutrons and protons were in equilibrium, so that (7.8) was satisfied.

Now we must ask when the relation (7.8) ceased to be satisfied. Since the neutron–proton mass difference is

$$Q = m_n - m_p = 1.23 \text{ MeV}, \tag{7.9}$$

it is clear that when the temperature dropped below a few MeV, the nucleons would no longer have enough thermal kinetic energy to remain in equilibrium and so have equal number densities. Clearly, since $Q \ll \{m_p, m_n\}$, their relative number densities would be given at this stage by the Boltzmann factor

$$\frac{n_n}{n_p} = e^{-Q/T}, \tag{7.10}$$

as long as they remained in statistical equilibrium because of the reaction rates. After the time when $T \sim 1\,\text{MeV} \Leftrightarrow t \sim 1\,\text{sec}$ at which $n_n/n_p \sim 1/4$, the remaining neutrons would simply undergo beta decay, resulting in a reduction of n_n/n_p, until the mean thermal energy dropped below the threshold for deuterium photodisintegration – that is, the energy at which there are a large number of thermal photons having sufficient energy to break up the deuterium nucleus.

7.3 The Helium Mass Fraction

Once deuterium can form, helium is rapidly built up to its equilibrium abundance. Intuitively, the reason for this is that deuterium is the nucleus most vulnerable to photodisintegration, while He^3, He^4, and tritium are relatively stable. Essentially all the deuterium is processed to He^4, because there are no sufficiently stable elements with atomic mass numbers from 5 to 8, so that no heavier elements are produced, and He^4 has the largest binding energy of those elements with mass number < 5. Small quantities of other light elements are produced at abundances $\ll 1$, mainly lithium, beryllium, and boron (Peebles 1966).

The physical reason why all the neutrons end up in helium nuclei is that the conditions in the universe make it overwhelmingly likely that they will be captured by protons. Since the neutron has no charge, the capture probability for a neutron on a proton is proportional to the geometrical cross-sectional area presented by the proton, $\sigma_p \sim r_p^2$ with r_p the nuclear radius (Silk 1980). The reaction rate will be proportional to the number density, n_n, of neutrons and their velocity, v_n. These we know to be $n_n \sim T^3$ and $v_n \simeq \sqrt{T/m_n}$. So if the capture reaction timescale

$$t_c \sim \frac{1}{r_p^2 n_n v_n} \ll \frac{1}{H} \tag{7.11}$$

then the expansion will have no damping effect on the nuclear reactions and they will run to completion. At the time of nucleosynthesis, one has $t_c H \sim 10^{-4}$, so this condition certainly holds. The rate of combination of deuterium nuclei into He^4 can be calculated in the same way, although

The Helium Mass Fraction

the capture cross-section is suppressed compared to the geometrical cross-section due to the electromagnetic repulsion between the proton charges. Allowing for this suppression, one finds that for the entire reaction

$$n + p \longrightarrow D + \gamma \qquad (7.12)$$

the capture timescale satisfies $t_c H \sim 10^{-3}$ (Silk 1980). The reaction that forms He^4 therefore proceeds rapidly to completion.

At this stage, we need to introduce some experimentally measured quantities: the binding energy of deuterium, which is about 100 keV, and the neutron lifetime, t_n, which is about 1000sec. Equation (6.24) allows us to estimate the time at which the temperature drops to the binding energy of the deuteron as $t(T = 100 \text{keV}) \simeq 300$sec. The neutron to proton ratio at T_f, the weak reaction freeze out temperature (the temperature at which the weak reaction rate drops well below the expansion rate) will have been reduced by beta decays at this time by an overall factor of $e^{-300/1000}$. The complete calculation shows that when the temperature drops below the deuterium binding energy, one has

$$T \sim 100 \, \text{keV} \implies \frac{n_n}{n_p} \simeq \frac{1}{7}. \qquad (7.13)$$

Since essentially *all* the neutrons are processed into He^4, with only traces remaining of deuterium, tritium, and He^3, the number density of He^4 nuclei is $n_{\text{He}^4} = n_n/2$. Because $n_n < n_p$ and the mass density of He^4 in atomic mass units is $4n_{\text{He}^4}$ and the total mass density is $n_n + n_p$, the resulting mass fraction of He^4 is generally

$$X(\text{He}^4) = \frac{2n_n}{n_n + n_p}. \qquad (7.14)$$

For the neutron to proton ratio (7.13) given by the calculation so far, this yields the value

$$X(\text{He}^4) \simeq \frac{1}{4}. \qquad (7.15)$$

To see this, note that equation (7.13) says that 1 nucleon in 8 is a neutron, so from every 16 nucleons we can put 4 into He^4, giving the mass fraction above.

When the temperature drops below about 10 keV, the mean nucleon energy is too low to enable fusion, and the nuclear reactions stop abruptly, freezing the He^4 abundance at the equilibrium value given by (7.15), and leaving miniscule traces of deuterium, tritium, He^3, and Li^4.

The value given by (7.15) is in remarkably good agreement with observations, and is relatively insensitive to the exact choice of parameters. Detailed research on cosmological element abundances is therefore more concerned with giving precise predictions for the abundances of Li^4, He^3,

H^3 and H^2. These, unfortunately, are both theoretically sensitive and extremely difficult to observe with any accuracy.

To what, then, is the result (7.15) sensitive? It must be sensitive to the expansion rate, since that determines the functional relationship $T(t)$, but it turns out that the reaction rates are much faster than the expansion rate during nucleosynthesis, unless we choose the energy density to be so large or small that it would be excluded by present-day observations. (Recall that we have a good limit of $0.01 < \Omega_{now} < 10$, and the final He^4 abundance only varies fractionally over this range of densities.) One factor which does crucially affect the final result (7.15) is the value (7.13) of the n/p ratio at the onset of nucleosynthesis. Recall that this ratio is determined by the number of neutrons that decay between the time when the weak reactions freeze and the onset of nucleosynthesis. Then the final He^4 abundance can vary if either

- the expansion rate is different, so that $t_n/t(T_f)$ varies, which would allow more or fewer neutrons to decay before the onset of nucleosynthesis, or
- the neutron decay lifetime is different from the value used in the computation.

Two factors are therefore important: in order to deal with the first adequately, we need to know the number of massless species present around the nucleosynthesis epoch, and to cope with the second we must know the neutron lifetime accurately. Again, unfortunately, at present the values of these quantities are too uncertain to permit any considerable improvement in the sharpness of the results already obtained. We find, then, that we must restrict ourselves to the statement that the picture we have built up provides a qualitatively simple and quantitatively interesting description of the process of primordial nucleosynthesis. Of course, not only theory is to blame here: the observations, and also their correct interpretation, are still under debate. The results of the preceding discussion should be sufficient to convince the reader that the theory we are developing of the evolution of the universe provides concord between our observations of the state of the universe at present and our theories of the universe's past.

7.4 Review

When the universe was sufficiently hot, at temperatures greater than the neutron mass, the weak interactions could maintain equilibrium between the neutrons and protons, which were therefore present in equal numbers. Once the temperature fell below this level, the weak interactions between nucleons could not maintain equilibrium, and the neutron number density fell steadily due to the decay of free neutrons. When the temperature fell below the deuterium nucleus photodissociation energy, the nucleon capture rate was still much higher than the expansion rate, so that all possible nucleon capture reactions proceeded to completion. Helium was formed by a proton and neutron first capturing each other to form a deuterium nucleus, and two deuterium nuclei then capturing each other to form a nucleus of helium. Although this process may leave traces of deuterium, helium-3, and lithium, no higher-mass elements are produced by this process, since there are no stable intermediate nuclei with masses just above that of helium-4. The helium mass fraction then resulting is limited by the available neutron fraction, since most of the neutrons were previously lost in decays. The helium mass fraction predicted on the basis of reasonable values of all the parameters is about one quarter, which is in good agreement with observations.

7.5 Exercises

Exercise 7.1 *The exact value of the final helium mass fraction $X(\text{He}^4)$ of around 25% is surprisingly sensitive to the values of other parameters involved in the calculation, such as the neutron lifetime and the neutron–proton mass difference. Quantify this sensitivity by calculating the ranges of these parameters for which $0.2 \leq X(\text{He}^4) \leq 0.3$.*

7.6 References

Hayashi, C. *Prog. Theor. Phys.* **5**, 224 (1950).

Peebles, P.J.E. *Astrophys. J.* **146**, 542 (1966).

Silk, J. *The Big Bang* (W.H. Freeman, San Francisco, 1980).

8

Cosmic Asymmetry and the Origin of Matter

Modern particle physics theories and cosmology first came together in the quest to explain the small residual asymmetry between matter and antimatter that accounts for all the matter now present in our universe. The result is a qualitative model that allows us to understand the quantity of matter remaining in the universe, after most of the matter and antimatter have annihilated, in terms of the microphysical asymmetry parameter of the particle model appropriate to our universe.

8.1 The Mystery of the Existence of Matter

One of the most interesting observations of modern cosmology is that the stuff of the universe consists entirely of matter, with very little antimatter present at all. Observations of the cosmic rays striking the upper atmosphere of the earth limit the intergalactic antiparticle to particle ratio to $\leq 10^{-4}$ at most. This is the sort of observation which contains a significant theoretical component, as we shall soon see. In particular, the lack of observations of very hard, penetrating cosmic gamma radiation from inside our galaxy is strong evidence that our galaxy at least is made entirely of matter. However, the cosmic ray evidence does not preclude the possibility that some galaxies are made entirely of antimatter. Instead, this possibility is excluded on causal grounds, by the fact that scales of galactic size are much greater than the horizon size before the start of nucleosynthesis. Obviously, any significant amount of annihilation after the era of nucleosynthesis would distort the helium mass fraction. The accuracy of the prediction for this mass fraction, obtained in the previous chapter, would lead us to expect that the generation of matter is completed well before the start of cosmic nucleosynthesis. This belief is supported by the existence of a straightforward model for the development of the cosmological matter–antimatter imbalance, which will be derived in this chapter.

8.2 Asymmetry as a Physical Theory

Probably the greatest triumph of unified field theories in the realm of cosmology is the prediction that the universe should consist entirely of matter, along with the accompanying quantitative relationship between the parameters of the unified theories and the present value of the baryon to photon ratio (Sakharov, 1967). This ratio is the simplest measure of the quantity of matter in the universe: clearly, the amounts of matter and antimatter respectively can be defined by the baryon and antibaryon number densities n_b and $n_{\bar{b}}$. We may then define the *net baryon number density*

$$n_B = n_b - n_{\bar{b}}. \tag{8.1}$$

Even more useful as a physical parameter, since it turns out to be very nearly constant throughout the history of the universe, is the baryon to photon ratio,

$$N = \frac{n_B}{n_\gamma}. \tag{8.2}$$

One can easily see that whenever particle number densities are conserved, N should remain constant, since $n_i \sim 1/V$ for all species i which are conserved. Photon numbers are also conserved by the expansion, since $T \sim 1/S$, so that $n_\gamma \sim T^3 \sim 1/S^3 \sim 1/V$ as before. The only alteration to this picture will occur when a particle species initially in thermal equilibrium drops out of thermal equilibrium and decays, thus enhancing the photon number n_γ. It is easy to see that such effects can only alter n_γ by factors close to unity.

In terms of N, the problem of the material content of the universe may be rephrased as the question of why $N \neq 0$. Observationally, $N \ll 1$, but $N > 0$ by the arguments already advanced above. This means that the universe may be considered to be in a state of *asymmetry* with respect to net baryon number – clearly, a symmetric state would show no preference for baryons over antibaryons, and so would have $N = 0$. Such symmetric states were once preferred theoretically, since they made unique theoretical predictions on the basis of their symmetries. They also led to a great deal of study of the possibility that equal numbers of galaxies are made of matter and antimatter, the general conclusion being that such models are implausible on the grounds of present evidence.

The generally held belief now is that our universe does not exist in a perfectly symmetric state, but rather that it arose from a state of symmetry, and then evolved dynamically into an asymmetric state. Accordingly, we believe that when the cosmic temperature was greater than some temperature T_X, there was zero net baryon number, $N = 0$ (Toussaint *et al.* 1979). The temperature T_X corresponds to the *unification scale*, the mass of the particle species which carry the unified force. Thus $T_X \sim m_X$, where such particles are called X-particles for convenience. Fairly general arguments show that $m_X \sim 10^{15}$ GeV, so we are here discussing dynamics that occurred long before nucleosynthesis ever began.

8.3 The Generation of Baryon Number

The main idea behind this model of baryon number generation is very simple indeed, and is as follows. Above T_X, all particles are effectively massless, and are hence in thermal equilibrium at $T > T_X$. The universe is thus in a symmetric state, since a thermal equilibrium state must be invariant under time reversal, and hence also invariant under the interchange of particles and antiparticles. Clearly, the net baryon number has to be zero in such a state. Where, then, can the baryon number come from? The answer is that the baryon number can be generated by the decays of the X-particles only as they fall out of equilibrium. This happens as the temperature falls through T_X due to the expansion (Toussaint et al., 1979; Weinberg, 1979).*

A final, but crucial, ingredient of the recipe for baryon number generation is that the gauge symmetry of the unified theory is broken below some energy scale – obviously, this will be $\sim m_X \sim T_X$. So when the X-particles decay, they can decay *asymmetrically* preferring end-products which have baryon number $n_B > 0$ over those with $n_B < 0$. Measuring the asymmetry by a parameter ϵ, the overall picture develops as follows:

1. $T > T_X$: The universe is in thermal equilibrium, with $n_B = 0$ and $n_X = n_\gamma$.

2. $T \sim T_X$: The gauge symmetry breaks, and the X-particles decay with asymmetry ϵ, leaving $n_b = (1 + \frac{1}{2}\epsilon)n_\gamma$ and $n_{\bar{b}} = (1 - \frac{1}{2}\epsilon)n_\gamma$, giving the net baryon density $n_B \simeq \epsilon n_\gamma$.

3. $T < T_X$: The universe remains in a state with $N \sim \epsilon$.

Clearly, the required asymmetry is $\epsilon \ll 1$, since observations presently show that $N_{now} \sim 10^{-9}$. This conclusion is independent of the exact details of any unified theory of particles that may eventually be developed (Kolb and Wolfram, 1980). Instead, it is now a *criterion* to be met by all theories, and by which all theories must be judged. It is an attractive feature of this model for baryon number generation that it requires only a tiny asymmetry in the decay amplitudes of the X-particles. We can thus see that our present universe, and all the structure it contains, is the final result of a miniscule defect in the symmetry of the very early universe.

---

* In unified theories, the underlying symmetry of the theory is expressed in its representation as a *non-abelian* symmetry group. Because this group is not abelian, the gauge particles – the particles carrying the forces of the theory – carry the same quantum numbers as the other particles present, so that the gauge particles are good candidates to be identified with the X-particles.

8.4 Review

In the very early universe, when the temperature is greater than the mass of the unknown gauge boson, these gauge bosons are in thermal equilibrium. As the temperature drops they annihilate, with a small asymmetry expressed as a preference for reactions which result in baryonic rather than antibaryonic products. When the gauge bosons have entirely annihilated, the resulting baryons and antibaryons also undergo mutual annihilation, leaving the small excess of baryons which have no antibaryons with which to annihilate. From this time on, the baryon to photon ratio of the universe remains roughly constant. Only a small asymmetry, much smaller than unity, is required to explain the baryon to photon ratio observed in our present universe.

8.5 Exercises

Exercise 8.1 *Assume that some number y of particle species decay through the evolution of the universe up to the present. Calculate and tabulate how the value, N, of the baryon to photon ratio changes over time.*

8.6 References

Kolb, E.W. and Wolfram, S. *Nucl. Phys. B* **172**, 224 (1980).

Sakharov, A.D. *Zh. Eksp. Teor. Fiz. Pisma Red.* **5**, 32 (1967).

Toussaint, D., Treiman, S.B., Wilczek, F. and Zee, A. *Phys. Rev. D* **19**, 1036 (1979).

Weinberg, S. *Phys. Rev. Lett.* **42**, 850 (1979).

9

Primordial Cosmological Inflation

This chapter is devoted to a presentation of the inflationary universe model, that is a model for start of the universe that attempts to address a number of cosmological problems, particularly the horizon problem, the flatness problem, and the problem of the origin of the perturbations which result in the formation of large scale structure. The inflationary model does this by introducing a scalar inflaton field (see section 9.3), whose energy density initially dominates the expansion of the universe. The special properties of this scalar field then result in a universe which begins by expanding exponentially, smoothing out all perturbations, driving the density parameter towards unity, and inflating the horizon to extremely large scales. At the end of the inflationary period, the scalar inflaton releases its energy as radiation, and the hot universe stage then begins. The fluctuations of the scalar inflaton during inflation ultimately become the perturbations which are imprinted on the universe after the end of inflation.

9.1 The Cosmological Problems

As we have described it in preceding chapters, the universe has become more easily accessible to our understanding: we have shown how the universe has expanded, what its thermal history is, how the light elements were synthesized, where the matter asymmetry comes from, and how structures may have begun to evolve. The sections presented so far comprise the standard cosmological model, but they still leave some questions unanswered. Why is Ω so close to unity now? We know from Chapter 5, especially equation (5.23), that $\Omega = 1$ is actually *unstable* – so why is the universe still so close to spatial flatness (the flatness problem)? Where do the primordial fluctuations come from which later evolve into galaxies (the perturbation problem)? And the most puzzling problem of all is the horizon problem already discussed in Chapters 2 and 4.

Recalling the earlier results of Chapter 4 in which we discussed horizons, we found there that if the scale factor $S \sim t^n$, then the particle horizon distance $d_p \sim t$. The horizon problem may be stated as follows. Suppose we set $S = d_p$ now, with $S \sim t^{2/3}$. Then if we follow the universe back in time, we find that d_p shrinks more rapidly than does S, so that the volume currently encompassed by our universe is rapidly broken up into a large number of *causally disjoint* regions as time goes backwards. Why is it, then, that distant regions of the universe appear remarkably similar to each other, when they have been causally disconnected for most of their history?

The simplest and most compelling answer so far known is that provided by cosmological inflationary models, which undergo a period of exponential expansion as described by equations (9.19) and (9.20). This turns out to be just what is necessary to solve the horizon problem. For example, consider the particle horizon distance, d_p, when $S(t) = S_0 e^{H_0 t}$: by (4.9), this is of the form

$$d_p = e^{H t_f} \int_{t_i}^{t_f} dt\, e^{-Ht} = \frac{1}{H}\left[e^{H(t_f - t_i)} - 1\right], \qquad (9.1)$$

and so if $t_f \gg t_i$, then $d_p \simeq H^{-1} e^{H(t_f - t_i)}$. The particle horizon can thus become arbitrarily large if the universe can be made to expand exponentially for a sufficiently long time. The universe would then turn out to be, against all appearances, completely causally connected, and we could consequently relax in the secure knowledge that whatever physical processes have acted in one part of the universe have also had equal opportunity elsewhere. The important thing about scalar fields is that they allow us to construct good physical models which produce just such an expansion history as would result in (9.1). We now show how such models can be implemented.

The central idea of inflationary models is that the energy density of the universe was not dominated by thermal radiation at the very earliest times, but by a different kind of field – normally a *scalar field*, which was not necessarily in thermal equilibrium with the radiation at the earliest stages of the cosmic expansion. In order to build inflationary models using scalar fields, we need to develop an understanding of some of their properties. This is done in the next section.

9.2 Matter Fields in the Early Universe

Because of the important role they play in modern views of cosmology, we give here a brief description of scalar fields and their interaction with gravity. Physically, a scalar field is a function which, at every point in space and time, contains information about the energy and pressure caused by a particular kind of particle. In quantum mechanical terms, a scalar field is a function describing a species of *spinless boson*. Mathematically, a

Matter Fields in the Early Universe

cosmological scalar field is best described by the corresponding *Lagrangian function*. Denoting the scalar by ϕ, this is

$$\mathcal{L}(\phi) = S^3 \left[\frac{1}{2}\dot{\phi}^2 - V(\phi) \right] \tag{9.2}$$

for the case of the homogeneous isotropic universe we have been studying. Here $V(\phi)$ is the *potential energy density* of the scalar field. The Lagrangian is of the form $KE - PE$, with KE the kinetic energy of the field, and PE the potential energy. Clearly, the energy density carried by the scalar field is thus

$$\rho_\phi = \frac{1}{2}\dot{\phi}^2 + V(\phi). \tag{9.3}$$

The equation governing the behaviour of this scalar is the usual *Euler–Lagrange equation*

$$\frac{d}{dt}\left(\frac{\partial \mathcal{L}}{\partial \dot{\phi}}\right) - \frac{\partial \mathcal{L}}{\partial \phi} = 0. \tag{9.4}$$

Here we have $\partial \mathcal{L}/\partial \dot{\phi} = S^3 \dot{\phi}$, and $\partial \mathcal{L}/\partial \phi = -S^3 dV/d\phi$, so substituting in equation (9.4) and dividing through by the volume factor S^3, we find the dynamical equation for the scalar field to be

$$\ddot{\phi} + 3H\dot{\phi} + \frac{dV}{d\phi} = 0. \tag{9.5}$$

We can deduce another property of a scalar field which is conceptually useful in the cosmological context, namely the pressure exerted by the field. Taking the time derivative of equation (9.3),

$$\dot{\rho}_\phi = \left(\ddot{\phi} + \frac{dV}{d\phi}\right)\dot{\phi} = -3H\dot{\phi}^2. \tag{9.6}$$

The last equality follows directly from equation (9.5). But also, from the general energy conservation equation (3.15), the scalar must satisfy

$$\dot{\rho}_\phi = -3H(\rho_\phi + P_\phi), \tag{9.7}$$

so that

$$\rho_\phi + P_\phi = \dot{\phi}^2, \tag{9.8}$$

and from equation (9.3), we then find the pressure exerted by a scalar field to be

$$P_\phi = \frac{1}{2}\dot{\phi}^2 - V(\phi). \tag{9.9}$$

Notice that this relation shows the pressure to be $P_\phi = \mathcal{L}/S^3$, namely the *Lagrangian density* of the scalar field. It is clear from the above few lines that the Lagrangian equation of motion (9.5) really expresses only the fact that the scalar field conserves energy in exactly the same way as the dust, gas, and fluids we have already studied in earlier sections.

Because the scalar field has a well defined energy density and pressure, we can define an effective adiabatic index, γ_ϕ, for it in direct analogy with equation (5.6):

$$\gamma_\phi = \frac{\rho_\phi + P_\phi}{\rho_\phi} = \frac{\dot\phi^2}{\frac{1}{2}\dot\phi^2 + V(\phi)}. \qquad (9.10)$$

Clearly, γ_ϕ will not normally be constant, as were the indices used in previous discussions. It will only be constant in two extreme cases:

- $V(\phi) = 0 \Rightarrow \gamma_\phi = 2$;
- $\dot\phi = 0 \Rightarrow \gamma_\phi = 0$.

Generally, as ϕ varies, γ_ϕ will vary between these extremes.

This is probably the appropriate place for a short digression on the nature of scalar fields. Almost everything in the universe is actually in the form of particles. However, it is quite often simpler to avoid dealing simultaneously with a whole multitude of particles, and to replace them with a continuous distribution which shares the same bulk properties, such as energy, pressure, and so on. This is exactly the approach we adopt when we describe the cosmic radiation – which really consists of a vast number of individual photons – as a radiation gas with energy density aT^4, pressure $aT^4/3$, and so forth. The scalar fields we are presently discussing play a similar sort of role, differing in that the particles which they represent are scalar bosons, and they generally have a nonzero mass. Photons, by contrast, are vector bosons, and are massless. Studying cosmological scalar fields is justified for a number of reasons: one is that it will usually be interesting to know how the universe behaves when its expansion is dominated by forms of matter other than conventional dust or radiation; another is that scalar fields are an integral part of all modern theories of elementary particles; and yet another is that they allow us to construct physically simple but conceptually powerful inflationary models.

Let us now consider the two extreme cases discussed after equation (9.10) in more detail. We shall restrict attention to models with $k = 0$ for the present. Also, for the rest of this chapter, we shall make the substitution $G = 1/M_P^2$ for the gravitational constant, where M_P is the Planck mass.

Purely kinetic energy

$$\gamma_\phi = 2 \iff V(\phi) = 0 \Longrightarrow \rho_\phi = \frac{1}{2}\dot\phi^2; P_\phi = \frac{1}{2}\dot\phi^2. \qquad (9.11)$$

The equations corresponding to (3.13) and (3.15) for this case are thus

$$3H^2 = \frac{4\pi}{M_P^2}\dot\phi^2; \qquad (9.12)$$

$$\ddot\phi + 3H\dot\phi = 0. \qquad (9.13)$$

The Chaotic Inflationary Model

We begin the solution process by separating (9.13) into

$$\frac{\ddot{\phi}}{\dot{\phi}} = -3\frac{\dot{S}}{S}, \qquad (9.14)$$

which gives

$$\dot{\phi} = \dot{\phi}_0 \left(\frac{S_0}{S}\right)^3. \qquad (9.15)$$

We are solving the system subject to the initial condition that at $t = t_0$, we have $\phi = \phi_0$, $S = S_0$ and so on. Substituting (9.15) into (9.12), we get

$$S^2 \dot{S} = \sqrt{\frac{\pi}{3}} \left(\frac{2\dot{\phi}_0 S_0^3}{M_P}\right) \qquad (9.16)$$

which has the solution

$$S(t) = S_0 \left[\frac{2\sqrt{3\pi}}{M_P} \dot{\phi}_0 (t - t_0) + 1\right]^{1/3}. \qquad (9.17)$$

Putting (9.17) back into (9.15), we find the time dependence of ϕ to be

$$\phi(t) = \phi_0 + \frac{1}{\sqrt{12\pi}} \left(\frac{M_P}{\dot{\phi}_0}\right) \ln\left[1 + (t - t_0)\right]. \qquad (9.18)$$

Equations (9.17) and (9.18) describe the evolution of a universe dominated by a scalar field with a vanishing potential. Clearly, these equations will be a very good approximation in the case that $\frac{1}{2}\dot{\phi}^2 \gg V(\phi)$. We shall describe this case, or approximation, as the *velocity dominated* case in future, since here the energy density and pressure are dominated by the velocity of the scalar field as it rolls across its potential.

Purely potential energy

$$\gamma_\phi = 0 \iff \dot{\phi} = 0 \implies \rho_\phi = V(\phi); P_\phi = -V(\phi). \qquad (9.19)$$

The solution here is $\phi = \phi_0$, $V(\phi) = V(\phi_0) = V_0$, and thus

$$S(t) = S_0 e^{H_0 t}; \quad H_0 = \sqrt{\frac{8\pi V_0}{3M_P^2}}. \qquad (9.20)$$

A constant scalar field with a constant potential thus acts like a *cosmological constant* term, and makes the universe expand *exponentially*, instead of merely as a power of the time.

9.3 The Chaotic Inflationary Model

In this section, we describe in detail the simplest of all inflationary models. It is a version of the original chaotic inflation model proposed by

Linde (1983) which has been slightly simplified, without losing any of its qualitatively important properties (Madsen and Coles 1988). This model is called *chaotic inflation* because it is developed from the realization that the initial conditions for inflation may differ unpredictably throughout the universe.

The basic idea is the same as that mentioned at the end of the last section, but a certain degree of complexity must still be introduced in order to make a workable model. The problem we must anticipate is that if the universe simply expands exponentially, then the radiation temperature (which goes as $T \sim 1/S$, recall) is rapidly reduced to near zero, and small perturbations are also completely washed out, so that there is nothing left to grow into the large scale structures such as the galaxies and clusters which we see all around us. What is more, the universe does not appear to be expanding exponentially at the present. It is therefore necessary to construct a model which not only gives sufficient inflation to solve the horizon problem, but also exits gracefully into a radiation dominated era, so that our total cosmological picture can retain all the successes of the hot big bang model which we have developed in the preceding chapters.

With these considerations in mind, then, we shall make the following assumptions. Assume that the energy density of the universe in its very earliest stages is initially dominated by a single scalar field, having a potential $V(\phi) = \frac{1}{2}m^2\phi^2$, and that the field velocity is initially small, $\frac{1}{2}\dot\phi^2 \ll \frac{1}{2}m^2\phi^2$, but do not neglect the $\dot\phi$ term entirely – it is this last point which allows the universe to exit from the inflationary era. Here the constant m can be identified with the *particle mass* of the scalar field, that is, m is the mass of the individual particles described by the field. In this context, the scalar field is normally known as the *inflaton*. This observation is motivated by inserting the chosen form of the potential into the Lagrangian function (9.2). Since we are assuming $|\dot\phi|$ is small and slowly varying over the cosmological timescale $1/H$ (that is, $|\ddot\phi| \ll |\dot\phi|$), we can approximate the cosmological equations as

$$3H^2 \simeq \frac{4\pi}{M_P^2} m^2 \phi_0^2, \tag{9.21}$$

$$3H\dot\phi + m^2\phi \simeq 0. \tag{9.22}$$

Equation (9.21) gives

$$H \simeq H_0 \simeq \sqrt{\frac{4\pi}{3}} \frac{m\phi_0}{M_P} \Rightarrow S(t) \simeq S_0 \exp\left[\sqrt{\frac{4\pi}{3}} \frac{m\phi_0}{M_P} t\right]. \tag{9.23}$$

This solution will only be valid until $|\Delta\phi| \equiv |\phi - \phi_0|$ becomes comparable to $|\phi_0|$. To estimate when this happens, we use (9.23) to solve (9.22):

$$\frac{\dot\phi}{\phi} \simeq -\frac{m^2}{3H} \simeq -\frac{mM_P}{\sqrt{12\pi}\phi_0} \tag{9.24}$$

$$\Rightarrow \phi \simeq \phi_0 \exp\left[-\frac{mM_P t}{\sqrt{12\pi}\phi_0}\right]. \tag{9.25}$$

The exponential slow roll of ϕ down the slope of its potential, as seen from the solution (9.25), is then seen to have an *exponential relaxation time* t_e,

$$t_e \simeq \frac{\sqrt{12\pi}\phi_0}{mM_P}. \tag{9.26}$$

It seems a sufficiently good approximation to assume that for $t < t_e$, the *slow rolling* approximation described by (9.21) and (9.22) will be good. During this time, the universe inflates by a factor

$$n_e \equiv \frac{S_e}{S_0} \simeq \exp\left[4\pi\phi_0^2/M_P^2\right]. \tag{9.27}$$

From equation (9.27) it is evident that if $\phi_0^2 \gg M_P^2$, the inflationary factor can be an extremely large number. Note also that during the inflationary period, the energy density $\rho_\phi \simeq V(\phi) \simeq$ constant, so that, for example, the spatial curvature term, k/S^2, is reduced by a factor of n_e^2, which means that the cosmological density parameter Ω is driven towards unity, as would be expected from the fact that here $\gamma_\phi < 2/3$ throughout the inflationary period (see Figure 5.2). This is how the inflationary models solve the problem of cosmological flatness described in section 5.1. Similarly, from equation (9.1) we see that the horizon problem can be solved if the inflation lasts for a sufficiently long time, so that the present day size of the horizon can in fact be much greater than the naive estimate of $3t_{now}$ found in equation (4.10). So inflation can solve the horizon problem as well. In so doing, it establishes the visible universe as being a single, huge, causally connected volume. This ingredient provides the solution to most of the other cosmological problems as well. In this view of the universe, the cosmological principle must be reinterpreted as applying to our visible portion of the universe.

9.4 The Planck Limit

The analysis we have been doing so far is classical, in the sense that we have taken no account of quantum mechanical effects. This will be logical provided that quantum corrections are expected to be small. We have taken care only to consider matter contributions which can be described classically, but since we are now talking about energy scales approaching the Planck mass M_P, we need to be careful. For instance, quantum mechanical corrections will become important in gravitational interactions when particles of energy E interact via gravity and $E^2 G \sim 1$. Since $G = 1/M_P^2$, this means that an energy of order M_P is a limiting energy scale for using a classical description of gravity. Correspondingly, the shortest timescale we can consider classically is $t_P \equiv 1/M_P$, and all energy densities must be bounded by M_P^4.

Applying this limit to the inflaton energy density, $V(\phi_0) \leq M_P^4$, in the case that the inflaton potential $V(\phi) = \frac{1}{2}m^2\phi^2$, as used in our model above, we find that there is a maximum value for ϕ_0:

$$|\phi_0| \leq \frac{M_P^2}{m}. \tag{9.28}$$

How large do we need $|\phi_0|$ to be in order that inflation is still able to solve the cosmological problems? The best guess is given by considering the age of the universe now: $t_{now} \sim 10^{10}\text{yr} \sim 10^{17}\text{sec} \sim 10^{60} t_P$ (recall that $t_P \sim 10^{-43}\text{sec}$). So we need the inflationary factor to be at least $n_e \sim 10^{60}$. From (9.27), this requires $|\phi_0| \geq \sqrt{5}M_P$, and from (9.28), this can be achieved if the inflaton mass is sufficiently small, $m \leq \sqrt{5}M_P$. This is not an unreasonable requirement, so that on these grounds alone the chaotic inflationary models are not only attractive, but also plausible. In fact, as we shall soon see, realistic models produce a total amount of inflation well in excess of this minimum requirement.

9.5 Density Fluctuations Generated by Inflation

The other major success of inflationary models is their prediction for the magnitude of density fluctuations produced during the inflationary period. These are simply estimated as follows. The inflaton is really a quantum field, so that it will be expected to experience small quantum fluctuations, with $\delta\phi \simeq H$ during the inflation (Guth and Pi 1982, Hawking 1982, Linde 1983, Linde 1985). ($\delta\phi \gg H$ would be forbidden by energy considerations, since H is roughly the mean energy per particle at this stage.) It is therefore easy to estimate $\delta\rho/\rho$ thus. We have, during inflation,

$$\delta\rho \simeq m^2 \phi \, \delta\phi \simeq m^2 \phi H \; ; \; \rho \simeq \frac{1}{2} m^2 \phi^2 \; ; \; H \simeq \sqrt{\frac{4\pi}{3}} \frac{m\phi}{M_P}, \tag{9.29}$$

so that we find

$$\frac{\delta\rho}{\rho} \simeq \frac{2H}{\phi} \simeq 4\sqrt{\frac{\pi}{3}} \frac{m}{M_P} \Rightarrow \frac{\delta\rho}{\rho} \sim \frac{m}{M_P}. \tag{9.30}$$

Equation (9.30) is our principal result in this section. Notice that it does not depend on the exact value of ϕ_0, but only on the inflaton mass m. From (10.40), and the fact that the cosmic background radiation has temperature fluctuations $\delta T/T \simeq 10^{-5}$ on large scales, we see that if $m \simeq 10^{-5} M_P$, then our inflationary model can generate fluctuations which comply with the limits imposed by the observed structure of the cosmic background radiation. Notice also that there is no reference to any particular scale for $\delta\rho/\rho$, so that we would expect the fluctuations on all scales to be of comparable magnitudes if they are still well inside the linear regime. Fluctuations

Density Fluctuations Generated by Inflation

of this magnitude and with such a scale-independent amplitude spectrum are known to be in relatively good agreement with most successful models of structure formation (Harrison 1970, Zeldovich 1972). Another way of seeing this is to notice that during the inflationary epoch, the universe is in an effectively stationary state: the expansion rate, energy density, and horizon size are all constant, so that the universe does not appear to be evolving at all. In the absence of time dependence of physical quantities, the gravitational field is accurately described by the Poisson equation:

$$\nabla^2 \Phi = 4\pi G \rho, \tag{9.31}$$

where Φ is the gravitational potential (not to be confused with the inflaton scalar field ϕ). Dimensionally, this means that on scales comparable to the horizon size $l_{hor} \sim t_{age} \sim 1/H$,

$$\Phi \simeq \frac{4\pi G \rho}{H^2} \tag{9.32}$$

while on any other fixed scale $\lambda < 1/H$, the fluctuations in the gravitational potential are derived from equation (9.31) to be

$$\delta\Phi \simeq 4\pi G \rho \lambda^2. \tag{9.33}$$

Combining equations (9.32) and (9.33) gives the fractional gravitational potential perturbation on a scale λ as

$$\frac{\delta\Phi}{\Phi} \sim \frac{\delta\rho}{\rho}\left(\frac{\lambda^2}{t_{age}^2}\right) \sim H^2 \lambda^2 \frac{\delta\rho}{\rho}. \tag{9.34}$$

Now, since H is constant, and in a stationary state $\delta\Phi/\Phi$ must remain constant, this means that the density contrast must vary with length scale so as to keep the right hand side of equation (9.34) constant:

$$\frac{\delta\rho}{\rho} \sim \frac{1}{\lambda^2}. \tag{9.35}$$

This is the scale free spectrum promised above. It is also often written in terms of the number N of particles contained within the volume λ^3, so that

$$\frac{\delta\rho}{\rho} \sim N^{-2/3}. \tag{9.36}$$

The elegant argument used here to derive the form of the spectrum of the inflationary perturbations was first presented by Barrow (1988). An alternative derivation is given by Clutton-Brock (1993) where some other interesting results relating to inflationary perturbations are presented.

It is important to notice that the spectrum of perturbations produced by inflation is far smoother on large scales than would be expected from purely random statistical fluctuations. Equation (9.36) can also be written in terms of the number density n of particles, $\rho = mn$, where m is the inflaton mass. Then on a particular scale λ one has $\delta N/N = \delta n/n$, so that (9.36) becomes

$$\frac{\delta N}{N} \sim N^{-2/3}. \qquad (9.37)$$

On the other hand, a purely random statistical fluctuation of a number N produces a fluctuation $\delta N \sim \sqrt{N}$, so that

$$\left.\frac{\delta N}{N}\right|_{random} \sim N^{-1/2}. \qquad (9.38)$$

The necessary conclusion is that inflation very effectively smooths out fluctuations to well below their statistical amplitude.

9.6 The End of Inflation

We remarked earlier that, in order to be successful, an inflationary model must provide a method of ending the inflationary period, and returning the universe to a radiation dominated state, so that the remaining evolution of the universe will proceed as described in earlier chapters, with baryons being generated and light nuclei being synthesized according to the picture we have already built up. The single best feature of the inflationary models is that it solves some of the problems of the earlier cosmological model based on a hot big bang, while retaining all the successes of that model.

Equation (9.26) gives an estimate of the length of time for which the slow rolling approximation of equations (9.21) and (9.22) will be valid. At the end of that time, the inflaton will be in the fast rolling regime, and will rapidly reach the minimum of its potential, through which it will then oscillate rapidly. In so doing, it will radiate most of its energy as thermal radiation, causing the universe to *reheat*. Since the inflaton energy density will be velocity dominated at this stage, the maximum thermal energy density which can result from this process can be estimated in terms of a *reheat temperature*, T_R:

$$\frac{\pi^2}{15}T_R^4 \simeq \frac{1}{2}\left\langle \dot{\phi}^2 \right\rangle_{\phi=0}. \qquad (9.39)$$

The universe then enters a radiation dominated era whose further evolution has already been discussed in the earlier sections of this book.

9.7 Dissipation of Primordial Inhomogeneity

We have already explained how the process of inflation generates a scale free spectrum of linear density fluctuations which grow into the present day large scale structure. What happens to fluctuations which are present before the onset of inflation? The simple answer is that they are smoothed out by the rapidity of the expansion, since there are in this case no growing modes. As is discussed in detail in section 10.3 on the Jeans mass, pressure gradients only act to prevent the growth of density fluctuations. This means that to demonstrate the absence of growing modes we need only consider the pressureless case. From equation (10.15), we have an equation for the dependence of density fluctuations when $H \simeq$ constant:

$$\ddot{\delta} + 2H\dot{\delta} - \frac{3}{2}H^2\delta = 0. \tag{9.40}$$

Since the coefficients are constant, the solutions to this equation are easily found to be

$$\delta(t) \sim \delta_{\pm}(t_0)e^{\pm it/\sqrt{2}}e^{-Ht}, \tag{9.41}$$

so that the solutions are oscillating, but modulated by a strong decaying exponential factor. Comparison with equation (9.1) shows that pre-existing perturbations can be thought of as *sticking* to the horizon, and, thus, as being diluted by the enormous expansion factor. This discussion shows that inflation makes the universe more *homogeneous*, and thus goes a long way to explaining the otherwise mysterious large scale homogeneity of the universe.

9.8 Eternal Chaotic Inflation

In an earlier section, we gave an estimate for the magnitude of small density fluctuations generated by inflation. The simple estimates of equations (9.29) and (9.30) are valid when $\phi \simeq$ (few)M_P, and so describe the fluctuations which will be seen in our part of the universe quite well. We can see this as follows: recalling that the Planck length is $l_P \sim 10^{-33}$cm, we see that if our universe is about 10^{10}yr old, then the horizon size at present is $d_h \sim 10^{28}$cm $\sim e^{62}l_P$, so that all of our present universe is produced during the last stages of the slow roll of the inflaton – this last point can be seen directly by examination of (9.27). We now wish to address the interesting question of what happens on scales much larger than this, which correspond to $|\phi| \gg M_P$. Although such scales are inaccessible to our observations – and will be for an enormously long time to come – we are merely projecting the physics we have already introduced, and not inventing new equations (Linde 1986). The exercise may thus be seen as a sound form of speculation, with a physical basis. In order to describe the behaviour on such large scales, we will need to derive more accurate estimates of the behaviour of fluctuations when $|\phi| \gg M_P$, which is in itself a rewarding exercise.

The magical expression to derive is one relating the local values of the expansion factor with the inflaton velocity and the fluctuation amplitude. This is important, because $\delta\rho/\rho$ may be larger than unity on sufficiently large scales, so that we cannot assume that it, or $\dot{\phi}$, are dynamically negligible here. Consider the volume density of energy in any form in the inflationary universe. This may be the density of radiation, of gravitational waves, or anything else like that, so long as it will bear an imprint. Clearly, this will have

$$\rho \sim e^{qHt} \; ; \; q = \text{constant} \tag{9.42}$$

because of the exponential expansion. Now, if there are fluctuations in ρ, they will only be due to fluctuations in *either* of H or t, since we could always *reset* the local time in such a way that one of these is constant. Here we shall put $H = $ constant at any particular time. Neglecting factors of the constant q, we obtain

$$\frac{\delta\rho}{\rho} \sim H\,\delta t, \tag{9.43}$$

where δt is the assumed fluctuation in time which accounts for the fluctuation of ρ. Now, since these fluctuations are caused principally by variations in $\dot\phi$, we can estimate

$$\delta\phi \sim \dot\phi\,\delta t, \tag{9.44}$$

so that equation (9.43) becomes

$$\frac{\delta\rho}{\rho} \sim \frac{H\,\delta\phi}{\dot\phi} . \tag{9.45}$$

Since the fluctuations are expected to be dominated by those with $\delta\phi \sim H$, we can use this relation in a more useful form,

$$\frac{\delta\rho}{\rho} \sim \frac{H^2}{\dot\phi} . \tag{9.46}$$

In the model we have used in previous sections, with $V(\phi) = \frac{1}{2}m^2\phi^2$, we can derive the more general form of expression for $\delta\rho/\rho$: substituting $H^2 \sim V/M_P^2$, $\dot\phi \sim V'/H$, $V \sim V'\phi$, we find

$$\frac{\delta\rho}{\rho} \sim \frac{\sqrt{V}\phi}{M_P^3} \sim \frac{m}{M_P}\left(\frac{\phi}{M_P}\right)^2 . \tag{9.47}$$

In the region $\phi \sim M_P$, this agrees well with the simple estimate given in equation (9.30). However, it is clear that when $\phi/M_P \geq \sqrt{M_P/m}$, we will have $\delta\rho/\rho \geq 1$. Notice that the only restriction on ϕ was given by (9.28), so this zone of ϕ values clearly corresponds to a set of physically possible cases. Also, when $\sqrt{M_P/m} \leq \phi/M_P \leq (M_P/m)$, the fluctuations $\delta\rho$ will clearly dominate the energy density of the universe, since $\delta\rho > \rho$ here. We may thus describe the universe in this case as being *fluctuation dominated*. If this situation comes about, then in some regions of the universe, the fluctuations will eventually drive ϕ into the region where $\phi/M_P \leq \sqrt{M_P/m}$. These regions will undergo inflation as was described in the earlier sections. But there will still always be regions with the inflaton in the fluctuation dominated regime. Later on, some of these may enter the *friction dominated regime* where $\phi/M_P \leq \sqrt{M_P/m}$ and inflate as well.

Exercises

The general picture which emerges from these considerations is of a universe which is extremely inhomogeneous on very large scales (which are, however, many orders of magnitude greater than our own horizon) and consisting of many different domains, most of which will, by now, have experienced their own evolution through inflation and evolved galaxies. Perhaps there are also people – just like us in every essential way – who wonder what life is about, where they came from, and whether they are alone in the universe.

9.9 Review

At the Planck time, when the universe emerges into the classical era, the scalar inflaton has chaotic initial conditions, is most probably at a potential well away from zero, and dominates the energy density of the universe. It begins to roll down the potential very slowly, resulting in an exponential expansion of the universe, with the horizon being inflated to much larger scales at the present time than we would estimate on the basis of expansion as a power of time. Additionally, the density parameter is driven rapidly towards unity, and pre-inflationary fluctuations are smoothed out on all scales. The fluctuations of the scalar inflaton as it rolls down the potential are imprinted on the universe with an amplitude roughly that of the ratio between the inflaton mass and the Planck mass, and have a scale free (Harrison–Zeldovich) spectrum. When the inflaton reaches the bottom of its potential, it oscillates freely, losing its energy as radiation and reheating the universe. The hot phase then begins and continues as described in earlier chapters: the inflationary model therefore retains all the successes of the hot universe model.

Examining what happens in regions where the scalar inflaton has fluctuations large enough to drive it up the potential (the fluctuation dominated regime) rather than allowing it to roll slowly down (the friction dominated regime) leads one to conclude that there must be other parts of the universe where inflation is only just beginning, or where it has never yet occurred. The emerging picture is of the universe as a complex, chaotic and inhomogeneous one where predictable evolution (and the possibility of life emerging) can occur only in some regions.

9.10 Exercises

Exercise 9.1 *Estimate the inflaton mass m necessary to account for the magnitude of the observed small fluctuations in the cosmic background radiation at the present. For a range of masses about this value, draw a graph of the logarithm of the inflationary factor given by equation (9.27) against ϕ_0.*

Exercise 9.2 *By drawing on physical arguments of the kind used in this* chapter, derive plausible maximum values for the following parameters at the start of inflation:

1. The density parameter and hence the value of K;
2. The inhomogeneity in the inflaton field.

In doing so, keep in mind the Planck limit. How much inflation is required to reconcile these values with present day observations?

9.11 References

Barrow, J.D. *Quart. J. Roy. Astron. Soc.* **29**, 101 (1988).

Clutton-Brock, M. *Quart. J. Roy. Astron. Soc.* **34**, 411 (1993).

Guth, A.H. and Pi, S.-Y. *Phys. Rev. Lett.* **49**, 1110 (1982).

Harrison, E.R. *Phys. Rev. D* **1**, 2726 (1970).

Hawking, S.W. *Phys. Lett. B* **115**, 295 (1982).

Linde, A.D. *Phys. Lett. B* **129**, 177 (1983).

Linde, A.D. *Phys. Lett. B* **162**, 281 (1985).

Linde, A.D. *Phys. Lett. B* **175**, 395 (1986).

Madsen, M.S. and Coles, P. *Nucl. Phys. B* **298**, 701 (1988).

Zeldovich, Y.B. *Mon. Not. Roy. Astron. Soc.* **160**, 1P (1972).

10

The Evolution of Cosmic Structure

In earlier sections, we assumed that the universe could be well described as being homogeneous and isotropic about every point. We argued then that this was a very good approximation to the way that the material content of the cosmos is distributed on very large scales. However, on smaller scales, the universe exhibits a wealth of structure: galaxies, clusters of galaxies, and voids where there appear to be almost no galaxies at all. Since the galaxies are the sorts of place that harbour life – our galaxy does, at least – it is imperative that we try to understand how these magnificent and beautiful structures arise. The solution to this problem too, is one that can be at least partially understood within the model of the cosmos which we have so far developed in this work. However, a caution against excessive optimism is necessary. No complete theory of structure formation has yet been developed, although different phases of the formation process have been studied in great detail. We shall accordingly restrict ourselves to presenting those aspects of the theory which are broadly accepted as the basis for the physical process of structure formation, namely the theory of the growth of linear perturbations in an expanding universe.

10.1 Matter as a Fluid

We begin with the standard hydrodynamic equations for a self gravitating pressureless fluid. We then derive the equations which determine the behaviour of small perturbations in this fluid by perturbing this system and linearizing the resulting equations. We then insert the appropriate background solution for an expanding universe of the type already studied. By changing coordinates from ones fixed with the background (the *Euler picture*) to coordinates comoving with the fluid (the *Lagrange picture*) as we have done in earlier sections, we obtain solutions that show the existence of growing density perturbations. Next, we consider the effects which should be produced by the internal pressure of the fluid, and hence arrive at the extremely useful concept of the *Jeans mass*. Finally, we derive the relationship between the density perturbations of matter and the fluctuations observed in the cosmic radiation background temperature. The question of the origin of the perturbations discussed here was discussed in Chapter 9 on cosmological inflation.

To begin with, we shall neglect any pressure gradients that may be present in the fluid – the case with pressure gradients will be dealt with later on, in section 10.3. The basic hydrodynamical equations for a self gravitating fluid in this case are well known to be

$$\nabla^2 \Phi = 4\pi G \rho; \tag{10.1}$$

$$\frac{\partial \rho}{\partial \tau} + \nabla \cdot (\rho \boldsymbol{v}) = 0; \tag{10.2}$$

$$\frac{\partial \boldsymbol{v}}{\partial \tau} + (\boldsymbol{v} \cdot \nabla)\boldsymbol{v} = -\nabla \Phi. \tag{10.3}$$

Here Φ is the *Newtonian gravitational potential*, a generalized notation for the Newtonian potential for a point mass, $\Phi \sim 1/r$. Of the other quantities present, G is Newton's constant, ρ is the mass density of the fluid, and $\boldsymbol{v}$ is the fluid's local velocity vector. Note that all these quantities, apart from G, can depend on the time τ and the position vector $\boldsymbol{x}$. The notational use of τ for time here is necessitated by a later change of coordinates, but there is no physical difference between τ and the time t used in earlier sections. Focussing attention on these equations, we see that (10.1) says that the mass density is the source of the gravitational field, (10.2) is just the equation of continuity, expressing the conservation of mass, and (10.3) basically describes the balance of the forces acting on the fluid. The coordinate system $\{\tau, \boldsymbol{x}\}$ in which equations (10.1–10.3) hold is a rigid system with respect to which the fluid particles move – such a coordinate system is often referred to as a *laboratory frame*.

10.2 The Evolution of Small Perturbations

We shall now perturb (10.1–10.3) by assuming that the quantities $\rho, \Phi, \boldsymbol{v}$ are of the form $\rho = \bar{\rho} + \delta\rho$, $\Phi = \bar{\Phi} + \delta\Phi$, $\boldsymbol{v} = \bar{\boldsymbol{v}} + \delta\boldsymbol{v}$, where $\bar{\rho}, \bar{\Phi}, \bar{\boldsymbol{v}}$ are chosen to satisfy (10.1–10.3) by being the solutions corresponding to those of an isotropic and homogeneous expanding fluid, and $\delta\rho, \delta\Phi, \delta\boldsymbol{v} \ll \bar{\rho}, \bar{\Phi}, \bar{\boldsymbol{v}}$ respectively. In particular, by Hubble's law we always have $\boldsymbol{v} = H\boldsymbol{x}$ for the particle velocity relative to any choice of origin. Also, it is simpler, instead of $\delta\rho$, to work with the fractional perturbation in the density, $\delta \equiv \delta\rho/\rho \simeq \delta\rho/\bar{\rho}$, so that $\rho = \bar{\rho}(1 + \delta)$. Note that by the homogeneity of the cosmological model, there is no dependence on spatial position: $\bar{\rho} = \bar{\rho}(t)$ and $H = H(t)$ in the background solution. This is a reasonable model to choose, since we know from earlier sections that these conditions are good approximations to our universe.

The Evolution of Small Perturbations

Accordingly, we make the substitutions

$$\Phi = \bar{\Phi} + \xi \;;\; \rho = \bar{\rho}(1+\delta) \;;\; \boldsymbol{v} = H\boldsymbol{x} + \boldsymbol{u} \qquad (10.4)$$

in (10.1–10.3) and linearize the resulting equations about the solution $\bar{\rho}, \bar{\Phi}, \bar{v}$. We also drop the overbar from $\bar{\rho}$, since to the order of approximation we are using, $\rho \simeq \bar{\rho}$. The perturbation equations then turn out to be

$$\nabla^2 \xi = 4\pi G \rho \delta; \qquad (10.5)$$

$$\frac{\partial \delta}{\partial \tau} + H(\boldsymbol{x} \cdot \nabla)\delta = -\nabla \cdot \boldsymbol{u}; \qquad (10.6)$$

$$\frac{\partial \boldsymbol{u}}{\partial \tau} + H(\boldsymbol{x} \cdot \nabla)\boldsymbol{u} = -\nabla \xi - H\boldsymbol{u}. \qquad (10.7)$$

The best way to make further progress is to change to a set of coordinates which are comoving with the fluid – Lagrangian coordinates – by the transformation

$$\{\tau, \boldsymbol{x}\} \to \{t, \boldsymbol{r}\} \;:\; \tau = t, \; \boldsymbol{x} = S(t)\boldsymbol{r}, \qquad (10.8)$$

so that

$$\frac{\partial}{\partial t} = \frac{\partial \tau}{\partial t}\frac{\partial}{\partial \tau} + \frac{\partial \boldsymbol{x}}{\partial t}\frac{\partial}{\partial \boldsymbol{x}} \;;\; \frac{\partial \boldsymbol{x}}{\partial t} = \dot{S}\boldsymbol{r} = H\boldsymbol{x} \qquad (10.9)$$

which in turn implies

$$\frac{\partial}{\partial t} = \frac{\partial}{\partial \tau} + H\boldsymbol{x} \cdot \nabla \;;\; \boldsymbol{x} \cdot \frac{\partial}{\partial \boldsymbol{x}} = \boldsymbol{r} \cdot \frac{\partial}{\partial \boldsymbol{r}}, \qquad (10.10)$$

and also

$$\frac{\partial}{\partial \boldsymbol{x}} = \frac{1}{S}\frac{\partial}{\partial \boldsymbol{r}}. \qquad (10.11)$$

Here we have used the obvious symbolic notation $\nabla_{\boldsymbol{x}} = \partial/\partial \boldsymbol{x}$, and $\nabla_{\boldsymbol{r}} = \partial/\partial \boldsymbol{r}$. After these transformations, the equations (10.5–10.7) become

$$\nabla^2 \xi = 4\pi G \rho \delta S^2; \qquad (10.12)$$

$$\frac{\partial \delta}{\partial t} + \frac{1}{S}\nabla \boldsymbol{u} = 0; \qquad (10.13)$$

$$\frac{\partial \boldsymbol{u}}{\partial t} + H\boldsymbol{u} = -\frac{1}{S}\nabla \xi. \qquad (10.14)$$

This is the main system of equations which we shall use in our study. The single most important quantity appearing in these equations is the fractional density perturbation δ, often referred to as the *density contrast*. An evolution equation for δ can be extracted from the system (10.12–10.14) as follows. Take $(\partial/\partial t)(10.13)$ and $\nabla\cdot(10.14)$, then eliminate the terms containing $(\partial/\partial t)(\nabla \cdot \boldsymbol{u}) = \nabla \cdot (\partial \boldsymbol{u}/\partial t)$, then use equation (10.12) to eliminate $\nabla^2 \xi$. Finally, substitute (10.13) to convert any remaining terms like $H\nabla \cdot \boldsymbol{u}$ to terms in $\partial \delta/\partial t$. The resulting dynamical equation is

$$\frac{\partial^2 \delta}{\partial t^2} + 2H\frac{\partial \delta}{\partial t} - 4\pi G \rho \delta = 0. \qquad (10.15)$$

This equation was first derived in this way by Bonnor (1957).

Equation (10.15) is an equation for the evolution of (small) density perturbations in an expanding universe in the absence of pressure. There are two conceptually important special cases which can be easily analysed, and we pause to examine these before continuing with the theoretical development.

The first case is that of a stationary (that is, non-expanding) universe, which has $H = 0$. This was in fact the case originally investigated by Jeans (1928), before the expansion of the universe was discovered. In this case, the second term vanishes from equation (10.15), which then has solutions of the form

$$\delta(t, \boldsymbol{r}) = A_+(\boldsymbol{r}) \exp(\sqrt{4\pi G \rho}\, t) + A_-(\boldsymbol{r}) \exp(-\sqrt{4\pi G \rho}\, t), \qquad (10.16)$$

where $A_\pm(\boldsymbol{r})$ are arbitrary functions describing the spatial dependence of δ, and the background variables are all constant. The exponential growth of this solution led Jeans (1928) to conclude that the observed structures in the universe could have arisen from small, presumably thermal, fluctuations of the homogeneous background.

The second case is of more importance, because we know now that the universe is expanding. Notice that the form of (10.15) shows that the growth (and the decay) of δ is damped by the time-dependent factor $2H$. We take the background solution to be that corresponding to a $k = 0$ isotropic dust model. Then $S \sim t^{2/3} \to H = 2/3t$, and $4\pi G\rho = 3H^2/2 = 2/3t^2$. In this case, equation (10.15) reads

$$\frac{\partial^2 \delta}{\partial t^2} + \frac{4}{3t}\frac{\partial \delta}{\partial t} - \frac{2\delta}{3t^2} = 0. \qquad (10.17)$$

If we factor out the spatial dependence of δ by writing $\delta(t, \boldsymbol{r}) = \Delta(t)\sigma(\boldsymbol{r})$, then this reads

$$\frac{d^2\Delta}{dt^2} + \frac{4}{3t}\frac{d\Delta}{dt} - \frac{2\Delta}{3t^2} = 0. \qquad (10.18)$$

This is an *Euler equation*, and can thus be transformed into a linear differential equation with constant coefficients by the transformation $t = e^w$, so the equation and its solution turn out to be

$$\frac{d^2\Delta}{dw^2} + \frac{1}{3}\frac{d\Delta}{dw} - \frac{2\Delta}{3} = 0 \Rightarrow \Delta(t) = \Delta_+ e^{2w/3} + \Delta_- e^{-w}, \qquad (10.19)$$

with $\Delta_\pm$ being constants. The total solution for δ is thus

$$\delta(t, \boldsymbol{r}) = B_+(\boldsymbol{r}) t^{2/3} + B_-(\boldsymbol{r}) t^{-1}. \qquad (10.20)$$

The difference between the two cases just examined, as represented by the solutions (10.16) and (10.20), is a striking demonstration of the difference between a stationary and an expanding universe. The expansion of the universe damps the growth of density perturbations so effectively that the growth law is changed from an exponential to a power law form, the latter even having an exponent less than unity. As we shall see in the course of our further development on this topic, this change in the qualitative structure of the solutions leads to the reversal of Jeans' original conclusion: we no longer believe that purely linear growth of density fluctuations could have produced the observed large scale structure of our cosmos.

To see the force of this statement, one needs only to realize that the growth rate $\delta \sim t^{2/3}$ – clearly, the decaying mode $\delta \sim t^{-1}$ is irrelevant for the purposes of this discussion – is too slow for linear perturbations to have gone far into the nonlinear regime within the present age of the universe. Although we have not explicitly dealt with the growth and decay of fluctuations in the cases with $k = \pm 1$, in practical terms their solutions are not very different from that presented in equation (10.20), since we know that even at present the density parameter Ω is relatively close to unity. We restrict ourselves to the remark that when $k = \pm 1$, the equation for δ is still soluble, with the growing solution typically being like $\delta \sim S(t)$, so that if $k = +1$, the perturbations grow more rapidly, but have less time to grow, while if $k = -1$, the perturbations are growing more slowly than when $k = 0$, but have correspondingly more time in which to grow.

To return to the thread of our discussion on why linear fluctuations cannot, on their own, be responsible for the observed large scale structure, we make the point by asking how large linear density fluctuations could have grown in the time since the universe ceased to be radiation dominated: the anisotropies in the thermal cosmic radiation background suggest that at a redshift $Z \sim 10^4$, the density perturbations had a magnitude $\delta \sim 10^{-4}$. Since then, δ has grown roughly like $\delta \sim S \sim Z$, so that on this basis, we would expect that structures on most scales are going nonlinear, $\delta \geq 1$, only at the present. Since a typical galaxy represents a local density enhancement which is extremely nonlinear, $\delta \sim 10^6$, the linear growth model is clearly not sufficient on its own to explain the existence of large nonlinear structures on a range of scales in our present universe. There is as yet no completely satisfactory model of the nonlinear growth of perturbations, since this depends greatly on the details of the environment in which the perturbations must grow from the time at which they grow nonlinear. However, the linear growth model developed in this chapter provides a qualitatively good understanding of how small linear fluctuations produced in the very early stages of the universe can grow under the influence of gravity alone into the observed large scale structures.

10.3 Gravitational Energy and the Jeans Scale

The perturbation equations we have derived so far are valid in the case that the cosmic fluid is pressureless, an approximation which we believe to be well justified in the present state of the universe. However, there is strong evidence that this has not always been the case – for example, we would expect that the radiation pressure might have played a significant role in the early universe. How would the qualitative picture developed above be modified if the fluid did have an internal pressure? One expectation is that perturbations might be able to grow only until they became pressure supported – after all, this is how stars avoid collapsing all the way down to black holes. Does the same thing happen on cosmic scales? In fact, these expectations are borne out in a range of cases, but as usual, the universe manages to make up even more complicated and interesting physics for us than we might have expected.

The motivation for believing that a self-gravitating cloud of material could be pressure supported is easily derived if one thinks in terms of the ratio of its gravitational potential energy and its internal thermal energy (Weinberg 1983). The gravitational potential energy of a clump of density ρ, radius r and mass $M = 4\pi\rho r^3/3$ is like

$$E_P \sim -\frac{GM^2}{r}, \tag{10.21}$$

while the internal energy per unit volume is proportional to its internal pressure P,

$$E_I \sim Pr^3. \tag{10.22}$$

The internal pressure of the material should therefore be able to support it against its own gravitational force if

$$Pr^3 > \frac{GM^2}{r}. \tag{10.23}$$

Eliminating r from these relations, we see that the clump can remain pressure-supported provided that

$$M < M_J = \frac{P^{3/2}}{G^{3/2}\rho^2}. \tag{10.24}$$

This result will soon be derived directly by consideration of the dynamics of the fluid.

Now return to our fundamental equations (10.1–10.3). If pressure effects are included, then of these equations only (10.3) will require modification. Equation (10.1) will not, since the gravitational potential depends only on the mass density, and neither will (10.2) as long as mass is still conserved. Thus, we modify (10.3) to read

$$\frac{\partial \boldsymbol{v}}{\partial \tau} + \boldsymbol{v} \cdot \nabla \boldsymbol{v} = -\nabla \Phi - \frac{1}{\rho}\nabla P. \tag{10.25}$$

The new term $-\rho^{-1}\nabla P$ on the right of equation (10.25) is easy to understand on physical grounds. The negative sign is there for the same reason as that in $-\nabla\Phi$: the motion will tend towards regions of lower pressure. Also, pressure has dimensions of force/area, so ∇P is like a force/volume term, and since ρ =mass/volume, $\rho^{-1}\nabla P$ has dimensions of force/mass, which is the same as acceleration, so that it fits dimensionally in (10.25). Of course, $\rho^{-1}\nabla P$ is not in a particularly tractable form. However, we can improve matters somewhat by assuming that the fluid is *isothermal*, which should be a good approximation as long as no part of the fluid experiences large compressions. Since these would, in any case, correspond to having $\delta \gg 1$, we should be able to model the pressure of the fluid as being given by the isothermal gas law

$$P = nT = \frac{\rho}{m}T. \tag{10.26}$$

Here we are using units in which the Boltzmann constant is unity, n is the number density of the particles making up the fluid, and m is the mass of a single particle. The isothermal nature of the fluid gives $\nabla T = 0$ everywhere, so in the perturbed solution $P = \rho(1+\delta)T/m$. Recalling that $\rho = \rho(t)$, we find $-\rho^{-1}\nabla P = -(T/m)\nabla\delta$ for the contribution of the fluctuation to the pressure gradient. Equation (10.7) thus changes, in the presence of pressure, to

$$\frac{\partial \bm{u}}{\partial \tau} + H\bm{x}\cdot\nabla\bm{u} = -\nabla\xi - H\bm{u} - \frac{T}{m}\nabla\delta. \tag{10.27}$$

It is now a simple exercise to repeat the derivation of equation (10.15), using (10.27) in place of (10.7): equation (10.14) becomes

$$\frac{\partial \bm{u}}{\partial t} + H\bm{u} = -\frac{1}{S}\nabla\xi - \left(\frac{T}{m}\right)\left(\frac{1}{S}\right)\nabla\delta. \tag{10.28}$$

Finally, the general form of equation (10.15) turns out to be

$$\frac{\partial^2 \delta}{\partial t^2} + 2H\frac{\partial \delta}{\partial t} - \left(4\pi G\rho + \frac{T}{m}\frac{\nabla^2}{S^2}\right)\delta = 0 \tag{10.29}$$

in the case when the fluid has a non-negligible internal pressure.

In order to make equation (10.29) a little more useful, let us write $\delta(t,\bm{x})$ as a Fourier sum. Physically this corresponds to decomposing $\delta(t,\bm{x})$ into waves, all of which added together comprise the total density perturbation. The Fourier expansion is

$$\delta(t,\bm{x}) = \sum_k \delta_k(t)e^{i\bm{k}\cdot\bm{x}}. \tag{10.30}$$

Here the δ_k are the (time dependent) Fourier coefficients defining the strength of the contribution of the wave $e^{i\bm{k}\cdot\bm{x}}$ to $\delta(t,\bm{x})$, and $\bm{k}$ is the wavevector corresponding to oscillations of wavelength l where $|k| = 2\pi/l$. Substituting (10.30) into equation (10.29), we find that the modes separate: each mode $\delta_k(t)$ obeys (10.29) independently of the other modes. This clearly happens because (10.29) is a linear differential equation, and (10.30) is a linear decomposition of δ. The mode separation allows (10.29) to be written for every mode in the form

$$\ddot{\delta}_k + 2H\dot{\delta}_k - \left(4\pi G\rho - \frac{4\pi^2 T}{mS^2 l^2}\right)\delta_k = 0. \tag{10.31}$$

Further, the way that l and $\boldsymbol{k}$ are defined, they are clearly the *coordinate wavelength* and *wavenumber*. The physical wavelength corresponding to l is thus $\lambda = Sl$, since the coordinate wavelength must be scaled by the cosmic scale factor. We can therefore further improve (10.31) to

$$\ddot{\delta}_k + 2H\dot{\delta}_k + \left(\frac{4\pi^2 T}{m\lambda^2} - 4\pi G\rho\right)\delta_k = 0. \tag{10.32}$$

This equation governs the growth of each Fourier component of the density fluctuation separately when there is a pressure present in the fluid. Of course we could already have obtained the corresponding equation, without the pressure term, by the same method in the derivation of the previous section. There it was unnecessary however, since there were no spatial derivatives in the evolution equation for δ in the pressureless case.

There are two limiting forms of equation (10.32) which are of interest in helping us to understand the effects of pressure.

- If $4\pi G\rho \gg 4\pi^2 T/m\lambda^2$ then the typical solutions will be the same as those for the pressureless case, already displayed in the previous section, with growing and decaying solutions present. For example, each Fourier mode will grow with $\delta_k \sim S(t)$ separately. The dominant effect is thus the self gravity of the fluid.

- When $4\pi G\rho \ll 4\pi^2 T/m\lambda^2$, new behaviour arises, and the solutions to (10.32) are *oscillatory*.

These conclusions neglect the possible effects of the damping factor $2H$ caused by the expansion, but will be valid in the extreme regimes specified. They are both easier to visualize if we return to the case originally treated by Jeans, and neglect the effects of expansion for the moment, $H = 0$. Then the two limiting cases mentioned above give

$$H = 0;\ 4\pi G\rho \gg \frac{4\pi^2 T}{m\lambda^2};\ \ddot{\delta}_k - 4\pi G\rho\delta_k \simeq 0 \Longrightarrow \delta_k \sim e^{\pm qt}, \tag{10.33}$$

$$H = 0;\ 4\pi G\rho \ll \frac{4\pi^2 T}{m\lambda^2};\ \ddot{\delta}_k + \frac{4\pi^2 T}{m\lambda^2} \simeq 0 \Longrightarrow \delta_k \sim e^{\pm ipt}. \tag{10.34}$$

So we conclude that when $4\pi G\rho \ll 4\pi^2 T/m\lambda^2$, the mode δ_k corresponding to (coordinate) wavelength λ does not grow or decay, but *oscillates*, like an acoustic wave, which is also a compressional (density) disturbance. Since the gas temperature T should be the same as the radiation temperature, which is uniform in our treatment by our original assumption of isothermality, we conclude that at each period in time, there are wavelengths which are oscillating because they are pressure supported and hence unable to collapse. There must also be other (much longer) wavelengths which correspond to growing perturbations collapsing under their own gravitational pull because they are too big to remain pressure supported. The clearest dividing line between these two regimes is where $4\pi G\rho = 4\pi^2 T/m\lambda^2$, and this wavelength is known as the *Jeans length*,

$$\lambda_J = \sqrt{\frac{\pi T}{mG\rho}}. \tag{10.35}$$

Equally often encountered in the literature is the *Jeans mass*, which is the total mass contained inside a volume with typical dimension λ_J:

$$M_J = \lambda_J^3 \rho. \tag{10.36}$$

The results of the preceding paragraphs show that if we wish to make rough estimates relevant to the structures formed (and forming) in the universe, then it will normally suffice to have an estimate for the Jeans mass at a given epoch. From this, we can make (accurate) rough guesses about the scale on which linear structures can exist at that epoch.

For example, at the present time, the gas temperature is probably around 3 K, the same as the temperature of the cosmic background radiation. Recall that if $k = 0$, then $\Omega = 1$, so that $3H^2 = 8\pi G\rho$. Substituting the necessary values, we find that the Jeans length is presently extremely small, $\lambda_J(now) \sim 10^{12}$ cm (Peebles 1971). So at the present, almost any linear perturbation is able to grow.

We have now developed all the necessary theory for our study of structure formation. We shall have occasion to return to this theory again to obtain a firm prediction for the form of the linear density fluctuations produced in the very early universe. It will also allow us to relate the magnitude of these fluctuations to the *microphysical* parameters of the cosmological aspects examined in later chapters. Before proceeding to that topic, however, we will close this chapter with a discussion of the relationship between matter perturbations and fluctuations in the cosmic background radiation.

10.4 Fluctuations of the Background Radiation

It is clear from the discussion so far that the density fluctuations of the matter, $\delta\rho/\rho$, are responsible for material structure of the universe, such as galaxies, clusters, and superclusters. Naturally, our interest in the development of these structures is strong, since these seem to be necessary for our own evolution as a species. However, perturbations of the matter may be expected to leave their imprint in other ways as well. For example, we would expect that photons present in the cosmic background might be deflected by the gravitational field of the denser clumps of matter, so that the cosmic background radiation may now be able to tell us about the inhomogeneities which are present in the cosmological matter distribution. This is especially important if the *visible* light distribution does not trace the mass distribution accurately – and it certainly does not do so on small scales – since the cosmic background radiation can then give us information about the perturbations of the unseen *dark matter*.

It is easy to estimate the magnitude of the temperature fluctuations in the cosmic background radiation, $\delta T/T$, caused by the matter perturbation, $\delta\rho/\rho$. Firstly, from the fact that the photon number density $n_\gamma \sim T^3$, we have

$$\frac{\delta n_\gamma}{n_\gamma} \simeq 3\frac{\delta T}{T}. \qquad (10.37)$$

Secondly, we are assuming that the matter is composed of baryons of mass m, so that $\rho = mn_B$, where n_B is the baryon number density. Thus,

$$\frac{\delta\rho}{\rho} \simeq \frac{\delta n_B}{n_B}. \qquad (10.38)$$

Finally, as we showed earlier, in the chapter on baryon number generation, we have $n_B/n_\gamma \simeq constant$ throughout most of the history of the universe, so that

$$\frac{\delta n_B}{n_B} \simeq \frac{\delta n_\gamma}{n_\gamma}. \qquad (10.39)$$

Equations (10.37–10.39) together show that the matter and cosmic radiation background fluctuations are related by

$$\frac{\delta T}{T} \simeq \frac{1}{3}\frac{\delta\rho}{\rho}. \qquad (10.40)$$

This result will hold over scales that are smaller than the horizon size at decoupling. The physical content of (10.40) is simply that, whenever the photons and baryons are coupled, the regions of the universe that have higher than the average densities must also be hotter than the average temperature.

On larger scales, one must estimate the effects of the gravitational potential of a perturbation on the photons coming from that direction in the sky. Consider cosmic background radiation photons coming from one part of the sky where the density is ρ, the gravitational potential is Φ, and the temperature is T. The average photon in this sky region of the cosmic background radiation has energy $E \sim T$. Compare these with cosmic background radiation photons coming from a nearby region of the sky where the density is $\rho + \delta\rho$, the gravitational potential is $\Phi + \delta\Phi$, and the temperature is $T + \delta T$. The average photon in this sky region of the cosmic background radiation has energy $E + \delta E \sim T + \delta T$. Now by conservation of energy, the ratio between the energies of the two regions is

$$\frac{E+\delta E}{E} \sim \frac{T+\delta T}{T} \sim \frac{\Phi+\delta\Phi}{\Phi}, \qquad (10.41)$$

so that the temperature fluctuations can be expressed in terms of the fluctuations of the gravitational potential:

$$\frac{\delta T}{T} \sim \frac{\delta \Phi}{\Phi}. \qquad (10.42)$$

Using the previously derived expression (9.34), which relates the gravitational potential fluctuations to the fluctuations of the density, we obtain the result

$$\frac{\delta T}{T} \sim \frac{\delta \rho}{\rho}\left(\frac{\lambda}{t}\right)^2. \qquad (10.43)$$

This gives the relation between the temperature and density fluctuations in terms of the length scale, λ, of the fluctuation and the age, t, of the universe. The result (10.43) is in accordance with that derived by a detailed relativistic analysis of the effects of gravity totalled over the paths of light rays by Sachs and Wolfe (1967). Silk (1986) and Barrow (1989) provide useful explanations and simpler derivations of the effect. As is pointed out by Barrow (1989), the relation (9.4), together with the observed smoothness of the cosmic background radiation, provides the best evidence for the validity of the cosmological principle, since we can deduce that the universe is in fact isotropic to a high degree of accuracy.

The practical meaning of the results (10.40) and (10.43) is that we can estimate the magnitude of primordial linear density fluctuations in the matter by measuring the corresponding present day perturbations in the cosmic background radiation. Since the background radiation fluctuations can be measured on a variety of scales with a high degree of precision, this is the most reliable method for estimating the magnitude of the initial density fluctuations in the matter. Although full understanding of the growth of large scale structures requires more knowledge of the growth of perturbations in the nonlinear phases, the linear fluctuation model developed here shows that the linear growth of perturbations can be well understood in terms of the action of gravity in the expanding universe.

10.5 Review

The Newtonian theory of small fluctuations results in a tractable second-order differential equation for the density contrast of a perturbation. In a non-expanding universe, the growing mode grows exponentially, but when the universe is expanding, the expansion damps out the exponential growth and leaves perturbations growing as a power of time. When the matter has internal pressure, its perturbations are prevented from growing on all scales smaller than a limiting value known as the Jeans scale. Fluctuations of the matter will generally result in correspondingly small temperature fluctuations. The magnitude of the small matter fluctuations from which the observed present day structures grew can therfore be measured indirectly by measuring the amplitude of the fluctuations of the cosmic background radiation.

10.6 Exercises

Exercise 10.1 *By considering the appropriate discriminant for equation (10.32), derive more general forms of (10.33) and (10.34). For the case of energy density dominated by radiation, $\rho \sim T^4$, derive a relation between T and λ_J, and draw graphs of this relation for various values of the density parameter Ω.*

Exercise 10.2 *Produce a graph of the amplitude of the perturbations that are just going nonlinear at the present time, $\delta\rho/\rho|_{now} \sim 1$, as a function of the density parameter Ω.*

10.7 References

Barrow, J.D. *Quart. J. Roy. Astron. Soc.* **30**, 163 (1989).

Bonnor, W.B. *Mon. Not. Roy. Astron. Soc.* **117**, 104 (1957).

Jeans, J. *Astronomy and Cosmogony*, (Cambridge University Press, 1928).

Lifshitz, E.M. and Khalatnikov, I.M. *Adv. Phys.* **12**, 185 (1963).

Peebles, P.J.E. *Physical Cosmology*, (Princeton University Press, 1971).

Sachs, R.K. and Wolfe, A.M. *Astrophys. J.* **147**, 73 (1967).

Silk, J. *Can J. Phys.* **64**, 147 (1986).

Weinberg, S. *The First Three Minutes* (André Deutsch, London, 1977).

11
Dark Matter and Structure Formation

Evidence for the existence of dark matter in the universe comes from observations of the dynamics of galaxies and galaxy clusters. Although one may be inclined to dismiss the existence of invisible matter as an unprovable hypothesis, one should not be too prejudiced in favour only of those components of the universe that can be seen in the visible region of the spectrum. This chapter presents some of the evidence for dark matter and describes the roles that dark matter may play in the formation of structure within the framework of both hot and cold dark matter models.

11.1 Evidence for Dark Matter

Much of our knowledge about the universe is gained by the study of the distribution of visible matter in the form of galaxies. As was discussed in Chapter 2, the amount of visible – that is, luminous – matter in our universe seems to correspond to an average density $\bar{\rho} \sim 10^{-30} \text{g/cm}^3$. Put in terms of the density parameter, the best average density estimates give $\Omega_{now} \simeq 1/10$, an observation that gives rise to the flatness problem discussed in section 5.1. The solution to the flatness problem was inflation. As discussed in Chapter 9, one of the effects of inflation is to drive Ω very close to unity. Since inflation solves many of the standard cosmological problems, we need to take its predictions seriously. Being able to see only luminous matter corresponding to $\Omega \simeq 1/10$ then gives rise to the *dark matter problem*. This approach to the Ω problem generates a theoretical motivation for the dark matter problem.

In addition to the theoretical motivation, there are two compelling observational reasons for suspecting the existence of dark matter. The first comes from observations of galactic rotation curves and suggests the existence of haloes of dark matter surrounding the visible component of galaxies, while the second comes from the gravitational binding of galaxies into clusters and provides evidence for a large additional contribution to the mass of galaxy clusters.

A galactic rotation curve is a graph showing how the rotational velocity of the galactic disc varies with the radius. Rotation curves are obtained by measuring the Doppler shifts of light from different parts of the galaxy. The rotation curve for the Andromeda galaxy, M31, is shown in Figure 11.1. This rotation curve is typical of what is found for spiral galaxies. Observations show that spiral galaxies have rotation curves which climb steeply from the centre, reach a maximum, dip down and then flatten out. Typically, the rotation curve is flat for about the outer half of the visible radius of the galaxy, and remains flat all the way to the edge of the visible material. Now spiral galaxies are gravitationally bound, and the outer regions are kept from collapsing inwards by their angular momentum. Energy balance for a rotationally equilibrated system means that

$$\frac{GM}{r} \sim v_{rot}^2 \qquad (11.1)$$

where v_{rot} is the rotational velocity at radius r and M is the mass contained within that radius. In the region where the rotation curve is flat, $v_{rot} \simeq constant$, the potential must also remain constant so $M \sim r$, which implies that the density $\rho \sim 1/r^2$ in the outer regions. The need for dark matter to explain the flat rotation curves is now clear: as one looks further out toward the edge of a spiral galaxy, one sees the luminous matter thinning out until it fades away into darkness. However, the rotation curves show us that the mass is still growing linearly with radius at the point at which the luminous matter fades away! The obvious conclusion to draw is that spiral galaxies are embedded in massive dark haloes. These haloes each contain mass comparable to that in the luminous component of the galaxy. From Figure 11.1 the mass estimate for M31 including the dark halo is $M_{M31} \sim 5 \times 10^{10} M_\odot$.

A similar argument applied to galaxy clusters produces the second observational reason for suspecting the existence of dark matter. Galaxy clusters are gravitationally bound groups of galaxies, that is, their mutual gravitation has allowed them to break away from the cosmic expansion to form stable groups. We shall examine the closest such system. The Local Group of galaxies is dominated by the mass of M31 and our own Milky Way, which are approaching each other at a cosmically small velocity $v_{approach} \sim 100\,\text{km/s}$ as measured by the blueshift of M31 (Raine 1981). Being gravitationally bound, the separation, r, of each galaxy from the centre of mass of the system is given by the balance of force as

$$\frac{d^2 r}{dt^2} = -\frac{GM}{r^2}. \qquad (11.2)$$

The term $d^2 r/dt^2$ can be estimated as R/t_{age}^2, where $R \sim 10^3\,\text{kpc}$, and t_{age} is the age of the bound system, $t_{age} \sim t_{now} \sim 10^{10}$ yr. Then the total

The Silk Scale

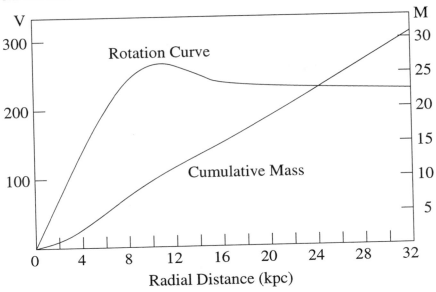

Figure 11.1 *The rotation and cumulative mass curves of the Andromeda galaxy M31. The scale on the left of the diagram shows measured rotational velocity in units of* km/s, *while that on the right shows the corresponding inferred cumulative mass in units of* 10^{10} $M_\odot$. *The levelling off of the rotational velocity at large distances is clear. After Silk (1980).*

system mass can be deduced as

$$M \sim \frac{R^3}{Gt_{now}^2} \sim 10^{12} M_\odot. \tag{11.3}$$

This mass is another order of magnitude greater than that derived from the flat rotation curve of M31, suggesting that galaxy clusters contain a diffuse component of nonluminous material contributing most of the mass of the cluster. The luminous matter appears from the empirical evidence to account for at most 10% of the mass of galaxy clusters. This finding shows that we need to take the possibility of the existence of dark matter very seriously indeed, because if it exists then most of the matter in the universe is dark, and many properties of the observed universe, such as the kind of structures that are formed in it, will be determined almost entirely by matter which we cannot observe directly.

11.2 The Silk Scale

The formation of structure consisting of baryons turns out to be greatly restricted during the period when the universe is opaque, namely before the radiation and the baryonic matter decouple. This is because density perturbations tend to contain increased densities of radiation as well. Before decoupling the matter and radiation are interacting strongly, and the pressure of the radiation inside a density perturbation can be large enough to hold the perturbation up under its own weight, so that it cannot begin to collapse until the matter decouples from the radiation.

The physical mechanism for this damping effect is the radiation pressure of the photon gas on the electrons in the matter plasma before recombination. The process is known as *Silk damping* after its discoverer (Silk 1968). The largest scale on which Silk damping is effective in preventing the formation of structure is known as the *Silk scale*. The Silk scale can be estimated as follows.

Consider a photon diffusing through the matter plasma at some time before recombination: the matter is totally ionized, and therefore consists of a dense plasma made up entirely of charged particles. The electrons in particular interact strongly with the radiation, so that the photon cannot travel in a straight line, but is constantly scattered by encounters with electrons. It therefore follows a random walk with an average step length equal to λ, the photon mean free path (see Stauffer and Aharony 1992). The mean free path must depend on the number density of electrons, n, and the electron–photon scattering cross-section, σ, known as the Thomson cross-section. Dimensionally, the relation between these quantities must be

$$\lambda = \frac{1}{n\sigma}. \tag{11.4}$$

Now the area, L^2, covered by a random walk is proportional to the time, t, over which the random walk elapses, with the proportionality given by the diffusion coefficient, D (Stauffer and Aharony 1992)

$$L^2 = Dt \tag{11.5}$$

and dimensionally it is easy to see that the diffusion coefficient is given by the mean free path, λ. Putting this together, the scale, L_s, up to which perturbations have been Silk damped at a time t is given by

$$L_s \sim \sqrt{\lambda t} \sim \sqrt{\lambda/H}. \tag{11.6}$$

Put another way, the Silk scale at a time t, $L_s(t)$, is the geometric mean of the photon mean free path, $\lambda(t)$, and the size of the horizon, $1/H(t)$. Since the electrons and nuclei making up the plasma are tightly coupled by the electromagnetic force, the damping effect of photons scattering off the electrons effectively damps the growth of fluctuations in the entire plasma.

The Silk scale is most frequently expressed in terms of the mass contained within a volume of that scale, $M_s \sim \rho L_s^3$. Since the most important manifestation of Silk damping is the *largest* scale on which the growth of structure is damped, the *Silk mass* is defined as the mass contained within the Silk scale at the time of recombination. Substitution in the above relations gives

$$M_s \sim \frac{m_p t_{rec}^{3/2}}{\sqrt{n_{rec}} \sigma^3}. \qquad (11.7)$$

Typical estimates for these parameters at decoupling show that the Silk mass corresponds to the galaxy or galaxy cluster scale.

It is important to understand the difference between the Jeans and Silk scales. In terms of baryonic perturbations, the Jeans scale is defined by the internal pressure of the baryons themselves, and is the scale on which the pressure of the baryons can support themselves against collapse under their own self gravity. The Silk scale, on the other hand, is defined by the effective pressure (via the medium of the electrons) of the radiation on the baryons.

11.3 Hot Dark Matter

Hot dark matter is dark matter that is still in thermal equilibrium. The archetype of hot dark matter models is built on the basic idea that neutrinos may have small masses of the order of a few electron volts. The other property of neutrinos, that they are weakly coupled to all forms of matter, including themselves, and so are effectively collisionless, is also typical of hot dark matter models.

Recall that in the very hot stages of the early universe, the neutrinos were in thermal equilibrium with the radiation, so that their relative number densities have been effectively preserved throughout the expansion,

$$n_\nu \sim n_\gamma. \qquad (11.8)$$

However, despite their number density remaining roughly equal to that of the photons, the neutrinos have been decoupled from all other particles since the time of weak reaction freeze-out at a temperature of about a few MeV. The mass density in neutrinos is therefore roughly

$$\rho_\nu \sim m_\nu n_\nu \sim m_\nu n_\gamma \qquad (11.9)$$

and since the baryonic mass density

$$\rho_b \sim m_b n_b \sim N m_b n_\gamma, \qquad (11.10)$$

where N is the baryon to photon ratio, the neutrino mass component will dominate the mass density in baryons if $m_\nu > N m_b$. Since $N \sim 10^{-9}$, this means that neutrinos with $m_\nu > 1\,\mathrm{eV}$ will dominate the mass density of the universe. In this case, gravitationally bound invisible structures made of neutrinos will be the gravitationally dominant objects of which the universe consists, and the luminous structures which we observe directly must be interpreted as a secondary effect whose structure is determined by the underlying structures of neutrino perturbations. Then galaxies or clusters of galaxies (depending on the appropriate neutrino clustering mass scales) can be understood as having formed within the enhanced gravitational fields of these neutrino haloes.

Supposing for the moment that neutrinos are clustered into haloes on large scales, it is possible to derive estimates of the neutrino mass on the basis that neutrinos, being subject to the Pauli exclusion principle, can occupy only a finite phase space. Consider a neutrino cluster of total mass, M, and radius, r, consisting of neutrinos with average speed, v, and corresponding momentum, $p = m_\nu v$. These neutrinos occupy a volume fraction, of phase space

$$V \sim \int d^3p \int d^3x \sim p^3 r^3 \sim (m_\nu v r)^3. \tag{11.11}$$

The corresponding maximum value for the mass of the cluster is

$$M \sim m_\nu V \sim m_\nu^4 v^3 r^3. \tag{11.12}$$

Now this cluster is assumed to be gravitationally bound and in thermal equilibrium, so that the average neutrino kinetic energy is approximately the same as its gravitational binding energy within the cluster (a condition known as *virial equilibrium*). Thermal equilibrium gives $v \sim \sqrt{T_\nu/m_\nu}$, where T_ν is the neutrino temperature, $T_\nu \sim T_{rad}$, while gravitational binding gives

$$v^2 \sim \frac{GM}{r} \sim G m_\nu^4 v^3 r^2. \tag{11.13}$$

Eliminating the neutrino velocity gives the neutrino mass estimate in terms of the scale of the cluster and the neutrino temperature as

$$m_\nu \sim \left(Gr^2 \sqrt{T_\nu}\right)^{2/7}. \tag{11.14}$$

Assuming that gravitationally bound neutrino haloes exist on galactic or cluster scales results in neutrino mass estimates of a few eV. While this range of masses is too small to have been reliably detected in terrestrial laboratories, it can clearly have significant cosmological consequences.

Understanding the detailed evolution of neutrino perturbations to form structures is beyond the scope of this volume, but it is possible to derive an extremely elegant estimate of the mass of the smallest structures that can form in a neutrino-dominated universe. As with the estimates of the Jeans and Silk scales, one tries to imagine a physical process which will effectively wash out any perturbations before they can grow. We then estimate the largest scale on which this process can operate effectively.

In the case of a universe dominated by neutrinos, a little thought will show that an overdense fluctuation cannot survive if the neutrinos, being collisionless, can stream out of the fluctuation. The reverse effect will wash out underdense fluctuations when it is effective. The phenomenon is appropriately called neutrino free streaming. It will be effective if the neutrino horizon-scale crossing time is less than the age of the universe. Since the neutrinos are in thermal equilibrium, their average velocity will again be

$$v_\nu \sim \sqrt{\frac{T_\nu}{m_\nu}} \qquad (11.15)$$

and the age of the universe, t, will be given by the horizon scale, $t \sim 1/H$. Clearly, neutrinos will be able to cross the horizon scale as long as their velocity is unity (that is, the speed of light), which is true until the temperature drops to

$$T \sim m_\nu. \qquad (11.16)$$

Since in this scenario the expansion is dominated by the neutrinos and these are behaving like radiation up to this time,

$$H^2 \sim \frac{T^4}{M_P^2}. \qquad (11.17)$$

Substituting $T \sim m_\nu$ and $H \sim 1/t$, we find that all fluctuations are wiped out by neutrino free streaming up to the free streaming time

$$t_{fs} \sim \frac{M_P}{m_\nu^2}. \qquad (11.18)$$

The horizon volume corresponding to this time is clearly

$$V_{hor}(t_{fs}) \sim t^3 \sim \frac{M_P^3}{m_\nu^6} \qquad (11.19)$$

and the mass density at this time is like $\rho_\nu(t_{fs}) \sim T_{fs}^4 \sim m_\nu^4$. It can therefore be seen that neutrino free streaming will wipe out all perturbations with masses less than the free streaming mass $M_{fs} = \rho_\nu(t_{fs})V_{hor}(t_{fs})$

$$M_{fs} \sim \frac{M_P^3}{m_\nu^2}. \qquad (11.20)$$

For neutrino masses within laboratory limits, this scale typically turns out to be comparable to the supercluster mass.

The prediction of erasure of all masses smaller than this scale is an important component of the massive neutrino model, and is characteristic of hot dark matter models. In particular, it leads to the conclusion that in a universe filled with hot dark matter, structure formation occurs in a *top-down* fashion: the largest structures are the first to form, since all structure on scales smaller than the free streaming scale is suppressed. Smaller scale structures therefore can form only by fragmentation of already existing larger structures. Luminous baryonic structures must then form by baryons falling into the gravitational potential of the resulting neutrino haloes. Since the baryons are effectively cold at that stage, they cannot remain pressure supported against the gravitational attraction of the neutrino structures, and are expected to collapse violently. This is known as the *pancake* model of galaxy formation, named after the observation that the large scale structures in this model will generically be collapsing at different rates along different directions (since the neutrinos of the halo are collisionless, there is no reason to expect them to isotropize into a spherical form), and so will probably be extremely flattened orthogonal to the direction of fastest collapse.

11.4 Cold Dark Matter

In contrast to the prediction of hot dark matter theories that structure should form 'top-down', theories of structure formation based on cold dark matter predict that structure should form *bottom-up*. The essential idea of the model is that if there exists a heavy particle species, which was originally produced in thermal equilibrium with the radiation, then it could be present in sufficiently large numbers that this species can dominate the cosmic mass density, since its contribution is like

$$\rho_{cdm} = m_{cdm} n_{cdm} \qquad (11.21)$$

where m_{cdm} and n_{cdm} are the mass and number density of the cold dark matter particle. The other essential idea is that the particle should be *cold*, that is, it should drop out of thermal equilibrium with the radiation early on, and become noninteracting with all other species present. In this case, there would be no internal pressure so that the Jeans mass for the cold dark matter would drop rapidly to zero. Since there is no equivalent of the Silk damping or free streaming processes for cold dark matter, perturbations on all scales inside the horizon would simultaneously start to grow. Because the cold dark matter particles can still interact with each other, they would also be able to dissipate their kinetic energy through collisions, and would therefore form compact tightly bound structures. The dominant scale of structures in cold dark matter models is determined entirely by the specific details of the model, particularly the spectrum of initial gravitational perturbations. In this picture, the compact clusters of dark matter provide gravitational seeds, around which the luminous baryonic matter clusters loosely at first, then settles smoothly to the centre of the dark matter halo through collisional damping of its kinetic energy – following in the footsteps of the cold dark matter itself.

Although the details of structure formation with cold dark matter are highly model-dependent, it is possible to derive some interesting properties of cold dark matter models. The following is the generic structure formation scenario (Blumenthal *et al.* 1984). Initially, at the time when the dark matter freezes out and becomes cold, the baryons and cold dark matter have the same fluctuation amplitudes $\delta_b \sim \delta_{cdm} \ll 1$. Although the cold dark matter model does not itself predict the form of these fluctuations, it is common to assume for the sake of theory that they have a scale-free spectrum, $\delta M/M \sim M^{-2/3}$, such as might be derived from a primordial inflationary phase.

The baryonic fluctuations are prevented from growing by Silk damping on mass scales $M < M_S$. The amount of mass inside a fluctuation scales like $\rho\delta \sim \Omega H^2 \delta$, so the cold dark matter fluctuations contain more mass, and will therefore be expected to grow more rapidly, because

$$\Omega_{cdm}\delta_{cdm} \gg \Omega_b \delta_b. \tag{11.22}$$

In fact, cold dark matter fluctuations that are smaller than the horizon scale are prevented from growing by the kinematic damping effect as long as the expansion remains dominated by the radiation (Meszaros 1975). This is because the expansion timescale is

$$t_{expand} \sim \frac{1}{H} \sim \frac{1}{\sqrt{G\rho_{rad}}}, \tag{11.23}$$

while the timescale on which cold dark matter fluctuations can grow is

$$t_{grow} \sim \frac{1}{\sqrt{G\rho_{cdm}}}. \tag{11.24}$$

Comparing these timescales, we can see that cold dark matter fluctuations will not be able to grow, and will instead *stagnate*, as long as $\rho_{cdm} < \rho_{rad}$. In order to estimate the scales on which cold dark matter fluctuations will be able to grow, notice that when the dark matter became cold at a temperature $T_{freeze} \sim m_{cdm}$, the cold dark matter made up a small contribution to the energy density, most of which was still left in particle species which were still massless, and therefore behaving like radiation. Calling this energy density fraction $(\rho_{cdm}/\rho_{rad})_{freeze}$, and noticing that after this time, $\rho_{rad} \sim T^4$ while $\rho_{cdm} \sim T^3$, the radiation will cease to dominate the expansion at a cutoff temperature

$$T_{cut} \sim \left(\frac{\rho_{cdm}}{\rho_{rad}}\right)_{freeze} T_{freeze}. \tag{11.25}$$

Since $T_{freeze} \sim m_{cdm}$, this can be rewritten

$$T_{cut} \sim \frac{m_{cdm}}{g^*_{rad}}, \tag{11.26}$$

where g^*_{rad} is the average number of effectively massless particle species between T_{freeze} and T_{cut}, to be derived from the appropriate particle theory which gave rise to the cold dark matter candidate particle. It will generally be a good enough approximation to use the value of g^* evaluated at T_{freeze}. Once the universe has ceased to be radiation dominated, the horizon scale $\lambda_{hor} \sim 1/H \sim Z^{-3/2}$, so that the cutoff scale below which no cold dark matter fluctuations can have grown at the end of the radiation domination can be expressed in terms of the present horizon scale as

$$\lambda_{hor}(T_{cut}) \sim \lambda_{hor}(T_{now})Z_{cut}^{-3/2}. \qquad (11.27)$$

Since $Z_{cut} \sim T_{cut}/T_{now}$, the final expression for λ_{cut} becomes

$$\lambda_{hor}(T_{cut}) \sim \lambda_{hor}(T_{now})\left[\frac{g^*_{freeze}T_{now}}{m_{cdm}}\right]^{3/2}. \qquad (11.28)$$

The relation (11.28) shows that in cold dark matter models, the smallest scales on which dark matter structures can form are determined by the microphysical parameters of the model.

After the dark matter comes to dominate the expansion, the fluctuations δ_{cdm} on all scales larger than the cutoff scale continue to grow, so that $\delta_{cdm} > \delta_b$. However, after recombination, Silk damping no longer prevents the growth of small scale baryon perturbations, and

$$\delta_b \longrightarrow \delta_{cdm}. \qquad (11.29)$$

This happens on mass scales $M > M_{Jb}$, where M_{Jb} is the baryonic Jeans mass:

$$M_{Jb} \sim \rho_b \lambda_{Jb}^3. \qquad (11.30)$$

Here, λ_{Jb} is the baryonic Jeans length, $\lambda_{Jb} \sim \sqrt{T_b/G\rho m_b}$, and it should also be noted that here, $\rho = \rho_b + \rho_{cdm}$ is the total mass density (reviewing the derivation of equation (10.35) from equation (10.27) should make this distinction clear). The baryonic Jeans mass can then be written as

$$M_{Jb} \sim \frac{\Omega_b T_b^{3/2}}{G H \Omega^{3/2} m_b^{3/2}}. \qquad (11.31)$$

On scales smaller than M_{Jb}, the pressure of the baryonic gas prevents its density fluctuations tracking those of the cold dark matter. The baryon temperature, T_b, remains close to the radiation temperature, T_{rad}, as long as there is a significant number of residual free electrons to interact with the radiation, but some time after recombination (by the time that T is about an order of magnitude below T_{rec}) this mechanism eventually becomes ineffective, and the baryons also become cold.

The main feature of the cold dark matter model is that it makes possible the rapid formation of baryonic structures that would not be formed in the absence of cold dark matter. This can be seen most clearly from the baryonic Jeans mass: the denominator on the right of equation (11.31) is much larger than if only baryons were present by a factor Ω/Ω_b, and the baryonic Jeans mass is correspondingly reduced by a factor $(\Omega/\Omega_b)^{3/2}$. This also shows why it is essential that the dark matter is cold in these models: the dark matter does not contribute to the internal pressure of matter, and so the growth of structure is not limited by a large value of the Jeans mass. The luminous baryonic structures now visible in the universe are

Review

interpreted, in the cold dark matter model, as being merely the luminous tracers of an underlying dark structure consisting principally of cold dark matter particles.

A more specific model of structure formation with cold dark matter in the form of scalar bosons is examined in section 12.3 below, where we also show how to estimate the dominant collapse scale.

The overall picture of galaxy formation in the cold dark matter model is that the dark matter perturbation spectrum at the time of recombination depends on the microphysical parameters of the model for the generation of those perturbations. After the end of recombination, when the baryonic perturbations are no longer Silk damped, the baryon perturbations are driven to the same amplitude as the cold dark matter perturbations. Small mass fluctuations grow rapidly into bound structures, which then cluster hierarchically into successively larger gravitationally bound structures. Ultimately, the baryonic matter in systems of galactic mass or larger cools and condenses inside the dark matter haloes to form galaxies and clusters.

11.5 Review

Observations of galaxies and clusters suggest the need to take account of the possible existence of dark matter which contributes a significant quantity of the mass density of the known universe. Models of the effect of dark matter on large scales may be based on hot dark matter consisting of light particles with thermal velocities, such as massive neutrinos, or cold dark matter consisting of as yet unidentified massive particles. The structure formation scenarios resulting from these distinct model types are very different. Hot dark matter predicts top-down structure formation, with the first structures forming late and on the largest scale, determined by the time at which the neutrino free streaming scale becomes smaller than the horizon, and baryonic structures forming in the most dense regions of these enormous pancake structures. Cold dark matter predicts the smallest structures forming early on, first on the smallest scales and then accumulating into larger structures as these form, with baryons falling into the large density fluctuations in the cold dark matter. Neither dark matter model can produce a truly satisfactory picture of the formation of large scale structure when compared to the existing observations, but our study of structure formation has illustrated most of the important ideas, such as the interaction of matter and radiation, and the determination of significant collapse scales and epochs.

The important mass scale that determines structure scales depends on the material forming the structure: for cold dark matter it is the cutoff scale; for hot dark matter, such as massive neutrinos, it is the neutrino free streaming scale; and for baryonic matter, it is the Silk scale.

11.6 Exercises

Exercise 11.1 *Estimate the time of decoupling for $\Omega = \{0.1; 1; 10\}$. Calculate the corresponding Silk masses and classify them according to the size of present day structure that they represent.*

Exercise 11.2 *Tabulate the neutrino mass estimates obtained from considering scales corresponding to globular clusters; galaxies; clusters; superclusters, and present values of the neutrino temperature $T_\nu = \{\left(\frac{11}{4}\right)^{1/3}; 1\} \times T_{rad}$. Think of, and argue in favour of, a physical process by which the neutrino temperature could have been raised to equal with or even above the radiation temperature since the time of weak interaction freeze out.*

Exercise 11.3 *Using the neutrino masses estimated in exercise 11.2, estimate and classify the free streaming mass erasure scale M_{fs}. Try to find different sky maps, and see if you can detect any measurable scales on them. To what range of neutrino masses do these scales correspond?*

Exercise 11.4 *For a variety of cold dark matter models, plot the baryonic Jeans mass M_{Jb} as a function of redshift. Take canonical present values of the parameters to be $\Omega = 1$, $\Omega_b = 1/10$, and $H = 50\,\mathrm{km.s^{-1}.Mpc^{-1}}$ and explore nearby values of each. Track the baryon temperature by using the approximation that $T_b \simeq T_{rad}$ for $Z > 100$ and $T_b \simeq T_{rad}/(1+Z)$ for $Z < 100$ (Blumenthal et al. 1984). This is known to be a good approximation: derive a physical argument for the origin of the extra factor of $(1+Z)$ in the latter expression.*

Exercise 11.5 *Use the relation (11.28) to estimate the mass m_{cdm} required for the cold dark matter particle to form smallest structures corresponding to:*

1. *Globular clusters ($M \sim 10^5 M_\odot$).*
2. *Typical galaxies ($M \sim 10^{10} M_\odot$).*
3. *Superclusters ($M \sim 10^{15} M_\odot$).*

11.7 References

Blumenthal, G.R., Faber, S.M., Primack, J.R. and Rees, M.J. *Nature* **311**, 517 (1984).

Meszaros, P. *Astron. Astrophys.* **38**, 5 (1975).

Raine, D.J. *The Isotropic Universe* (Adam Hilger, Bristol, 1981).

Silk, J. *Astrophys J.* **151**, 459 (1968).

Silk, J. *The Big Bang* (W. H. Freeman, San Francisco, 1980).

Stauffer, D. and Aharony, A. *Percolation Theory, 2ed.* (Taylor and Francis, London, 1992).

12

Exotic Objects

This chapter is devoted to the study of objects that may have interesting cosmological consequences if they exist, but are at the same time outside the main stream of explanation of the physical properties of the universe presented in the rest of this book. Only a selection of such topics is possible. Those that have been chosen for presentation are: black holes (section 12.1), because they are so widely known in connection with the study of gravity; magnetic monopoles (section 12.2), because they give rise to an interesting, cosmological problem and the need to solve it by cosmic inflation; and boson stars (section 12.3), because they illustrate many of the ideas of gravitational stability explored in earlier chapters. These topics were also chosen because they can be presented using the same kind of simple physical reasoning which is the backbone of the approach of this book to cosmology.

12.1 Black Holes

Black holes are the mythical beasts of physics. General relativity predicts the existence of these objects as the end result of burnt out stars collapsing under their own gravity. However, the most interesting properties of black holes can be derived by simple physical arguments.

The original motivation for the idea of black holes comes from the observation that the speed of light is the maximum velocity anything can reach. To see how this leads to black holes, consider the velocity of escape, v_{esc}, from the surface of a mass, M, with radius, R. This is the velocity at which the kinetic energy of an object at the surface is just sufficient to allow it to escape from the gravity of the mass:

$$\frac{1}{2}mv_{esc}^2 = \frac{GMm}{R} \implies v_{esc}^2 = \frac{2GM}{R}. \tag{12.1}$$

Now think of what happens when, for a given mass, M say, the radius, R, is steadily reduced, as would happen when the mass was shrinking under its own gravity. Eventually the velocity needed for escape would become equal to the speed of light, $v_{esc} = c = 1$. This would happen at a radius

$$R = 2GM. \tag{12.2}$$

If the radius became any smaller than this, no object or particle could escape from the mass, which would have become a black hole. The radius $R = 2GM$ corresponding to the black hole radius of a given mass, M, is called the *Schwarzschild radius* of M after the person who found the first black hole solutions in general relativity. The sphere of radius $R = 2GM$ surrounding the black hole is called the black hole horizon, since it is the limiting surface beyond which an external observer cannot see.

An intriguing cosmological possibility is that black holes could have formed in the early universe. These are known as primordial black holes. Primordial black holes could contribute to the dark matter in large numbers if the fluctuations present in the early post-inflationary stages were large. An unattractive aspect of primordial black holes is that it is generally very difficult to devise an experiment to detect their presence, and even harder to discriminate them from other compact massive objects which by contrast may have radii much larger than their Schwarzschild radius. The existence of sufficiently small primordial black holes, however, is testable in principle, due to the surprising fact that black holes radiate. The discussion that follows draws on the ideas of McCrea (1986).

As is well known, the fractional energy of radiation near frequency, ν, inside any region is proportional to the frequency of the radiation in any direction, and there are three independent directions in space, so that

$$dE \sim \nu^3 \, d\nu. \tag{12.3}$$

Writing this in terms of the wavelength of the radiation, $\lambda = 1/\nu$

$$dE \sim \frac{d\lambda}{\lambda^5}. \tag{12.4}$$

This shows that the total energy in the radiation scales as $E \sim 1/\lambda^4$, and since we know that $E \sim T^4$, this says that the temperature of the radiation $T \sim 1/\lambda$ in accordance with what we found when considering the effect of the cosmic expansion on the radiation temperature.

Now imagine a black hole with horizon radius, r, as being filled with trapped radiation (this is strictly a thought experiment, since we have no way of examining the actual contents of a black hole). Then think of what will happen if the horizon shrinks by an amount dr: the black hole will lose an amount of energy

$$dE \sim \frac{dr}{r^5} \tag{12.5}$$

implying that there is a force

$$\frac{dE}{dr} \sim \frac{1}{r^5} \tag{12.6}$$

pushing inwards on the horizon. The interpretation of this is that thermal radiation escapes from the horizon of a black hole, doing work that moves the horizon inwards as it does so. As a consequence, the mass of the black hole is also gradually reduced, escaping as the thermal energy of the radiation. Most interestingly, the temperature of the radiation emitted from the black hole is easily seen to be

$$T_{bh} \sim \frac{1}{R} \implies T_{bh} \sim \frac{1}{GM}. \tag{12.7}$$

So the smaller a black hole becomes, the hotter it gets. This is obviously a runaway process: the immediate physical consequence is that black holes evaporate. The power radiated by a black hole will be proportional to the energy density of the radiation and the surface area of the black hole: $P \sim \rho_{rad} A_{bh}$ (Birrell and Davies 1982). We know that $\rho_{rad} \sim T^4$ and the area $A_{bh} \sim R^2 \sim (GM)^2$. The power radiated is the rate of energy loss by the black hole, and the only source for this energy is the mass of the black hole itself. Putting this together, we have a simple expression for the rate at which a black hole loses mass by thermal radiation:

$$\frac{dM}{dt} \sim T^4 (GM)^2 \sim \frac{1}{G^2 M^2}. \tag{12.8}$$

Since the temperature is inversely proportional to the mass, the black hole grows correspondingly hotter as it radiates. The rate of temperature increase is easily seen to be

$$\frac{dT}{dt} \sim GT^4 \implies T \sim (Gt)^{1/3}. \tag{12.9}$$

As the black hole radiates and grows hotter, it will eventually radiate all those particles which are in thermal equilibrium at its current temperature. A radiating black hole that starts radiating mainly in the radio frequency region will move through the electromagnetic spectrum to radiate X-rays and γ-rays. When $T \simeq 1\,\text{MeV}$, electron–positron pairs will start to be radiated as well. As the temperature rises, heavier particles will make their appearance in the radiation around the black hole. An interesting consequence of the radiation of particles is that if the black hole happens to start with an electromagnetic charge, it can radiate this away by producing an excess of particles carrying that charge. Similarly, if the black hole has angular momentum, it can lose that by radiating an excess of particles with spin aligned with the rotation of the black hole. Once a black hole reaches $T \simeq 1\,\text{MeV}$, at which it can begin to radiate particles other than photons, therefore, it rapidly loses any charge and rotation that it may have had initially.

How long does the evaporation process take? Integrating the relation (12.8), we find the time it takes for a black hole to evaporate entirely away,

$$t_{evap} \sim G^2 M^3. \tag{12.10}$$

Some feeling for this timescale comes from asking what would be the initial mass of primordial black holes that are completing their evaporation now. These have $t_{evap} \sim t_{age}$, so that their mass is about 10^{15} g and their radius is about 10^{-13} cm, or to put it another way, they are about the size of a hydrogen atom and weigh about the same as a medium sized mountain.

It is conceivable that the dark matter could be made of black holes. If they close the universe so that $\Omega = 1$, then $\rho_c = \rho_{bh} = n_{bh} M$, where n_{bh} is the number density of black holes. For example, if the critical density for closure is $\rho_c \sim 10^{-30}$ g/cm^3 and the black holes are of the right mass to be evaporating now, $M \sim 10^{15}$ g, then $n \sim 10^{-45}$/cm^3. Cosmically speaking, this is a small number, and there would be only a few thousand such black holes inside our solar system, probably too few to detect directly. Black holes at this mass are probably ruled out by the fact that they would leave an imprint on the cosmic background radiation as they evaporate. Black holes with masses much less than 10^{15} g would already have evaporated completely, while black holes with much larger masses would still be sufficiently cold that they would be undetectable except possibly by their gravitational effects.

12.2 Magnetic Monopoles and the Monopole Problem

Magnetic monopoles are point sources of the magnetic field corresponding to isolated north or south poles. Their existence is compatible with standard theories of electromagnetism, as was originally realized by Dirac. More recently, the existence of stable, heavy (within a few orders of magnitude of the Planck mass) magnetic monopoles was found to be predicted within the framework of unified particle theories. Despite this, monopoles did not become of interest in cosmology until Preskill (1979) pointed out that thermal production of monopoles in the early universe would lead to problems for the standard model of cosmology. This conclusion was based on the observation that since monopoles were thermally produced, there would be approximately one monopole per baryon, and that since $m_m \gg m_b$, $m_m \sim 10^{16} m_b$, the universe now should have a density parameter $\Omega \sim m_m/m_b \gg 1$. This poses such a serious problem for cosmology that it has come to be called simply *the monopole problem*.

The only obvious way around the monopole problem would be if, like other species, the monopoles and antimonopoles annihilated efficiently with each other. However, Preskill (1979) also demonstrated that monopole–antimonopole annihilation is in fact extremely *inefficient*. For the purpose of exposition, we shall follow the calculation of Turner (1982).

The most important step in modelling the annihilation of magnetic monopoles is to write down a differential equation for the rate of change of their number density, n_m. In the absence of any creation or annihilation process, the number density would merely be diluted by the expansion,

$\dot{n}_m + 3Hn_m = 0$ so that $n_m \sim S^{-3}$. Clearly, the annihilation of monopole pairs is proportional to n_m^2, while the creation of monopole pairs is proportional to the square of the photon number density, n_γ^2, and will also depend on the temperature because this will limit how much energy there is to be used in pair creation. Putting this all together,

$$\frac{dn_m}{dt} + 3Hn_m = -C\left[n_m^2 - f(T)n_\gamma^2\right]. \tag{12.11}$$

Here, C is the *collision rate factor*, and $f(T)$ is a function of cosmic temperature still to be determined. In order to determine this function, consider the case when the monopoles are in thermal equilibrium: then the collision term on the right hand side of equation (12.11) must vanish so that the monopole number density is diluted only by the expansion. This is known as the principle of detailed balance. Thus, $f_{eq}(T) = n_m^2/n_\gamma^2$. But, in thermal equilibrium the number density of particles with momentum, p, in the interval $(p, p+dp)$ is given by the phase space expression

$$n(p)dp = 4\pi p^2 dp \exp\left[E(p)/T\right] \tag{12.12}$$

and the total monopole number density is therefore found by integrating over all momenta (Mandl 1988)

$$n_m \sim \int_0^\infty dp\, p^2 \exp(-E/T). \tag{12.13}$$

Since the monopole energy consists of its rest energy, m, and its kinetic energy, $\frac{1}{2}mv^2$, this can be written

$$n_m \sim \int_0^\infty dp\, p^2 \exp(-m/T) \exp(-mv^2/2T). \tag{12.14}$$

Transforming this to a velocity integral by replacing the element of momentum space, $d(p^3) \sim p^2 dp$, by the corresponding element of velocity space, $m^3 d(v^3) \sim m^3 d^3v$, we have

$$n_m \sim m^3 \exp(-m/T) \left(\int_{-\infty}^{+\infty} dv \exp(-mv^2/2T)\right)^3. \tag{12.15}$$

The integral is a standard item, so this expression can be evaluated to yield

$$n_m \sim (mT)^{3/2} \exp(-m/T). \tag{12.16}$$

Since the photon number density $n_\gamma \sim T^3$, the monopole to photon ratio is finally given as

$$\left(\frac{n_m}{n_\gamma}\right)_{eq} = \left(\frac{m}{T}\right)^{3/2} \exp(-m/T). \tag{12.17}$$

Now the final important step is to notice that the production of monopoles by collisions of photons (or other species in equilibrium with the radiation) depends only on the photons being in equilibrium, regardless of whether the final monopole products are. So (12.17) is always the form of the monopole to photon ratio

$$\left(\frac{n_m}{n_\gamma}\right) = \left(\frac{m}{T}\right)^{3/2} \exp(-m/T), \tag{12.18}$$

and the creation coefficient function can always be written as

$$f(T) = \left(\frac{m}{T}\right)^3 \exp(-2m/T). \tag{12.19}$$

Now we need to model the collision rate factor C. Clearly, the collision rate per monopole depends on three components: the average cross section, σ, presented by an individual monopole, the average speed, v, with which other monopoles approach it, and the number, n_a, of possible channels for the annihilation reaction to produce other particles. Dimensionally, the velocity averaged cross section, $\langle \sigma v \rangle$, is like an area, and it is expected to depend on the average thermal energy, $E_T = T$, per collision so that $\langle \sigma v \rangle \sim 1/T^2$ (Preskill 1979, Turner 1982). Then the collision rate factor is

$$C \simeq \frac{n_a}{T^2}. \tag{12.20}$$

Putting this all together, the equation for the evolution of the magnetic monopole number density is

$$\frac{dn_m}{dt} + 3Hn_m = -\frac{n_a}{T^2}\left[n_m^2 - (mT)^3 \exp(-2m/T)\right]. \tag{12.21}$$

This equation can be solved together with the other evolution equations. The total cosmic energy density will be $\rho = gaT^4$, where g is the number of particle species that are behaving like radiation, namely those with masses $\ll T$. The radiation temperature is determined by the relation derived earlier that $\dot{T}/T = -\dot{S}/S$, and the scale factor is found from

$$\frac{\dot{S}}{S} = \sqrt{\frac{8\pi^3 g}{90 M_P^2}} T^2, \tag{12.22}$$

where we have substituted for the blackbody constant in these units, $a/2 = \pi^2/30$.

Now when the production and annihilation rates of monopoles are large compared to the expansion rate, H, the monopoles will have their equilibrium abundance (Turner 1982). This situation holds when the parameter $\Gamma = (n_m)_{eq} C > H$. If initially, at a temperature, T_i, we have $\Gamma(T_i) > H(T_i)$, then the monopole number density will remain at its equilibrium value and will track that value until it freezes out at a temperature, T_f, where $\Gamma(T_f) = H(T_f)$. The freeze out temperature is therefore found by solving the transcendental equation

$$\sqrt{\frac{90 n_a^2}{8\pi^3 g} \frac{M_P}{m}} \left(\frac{m}{T}\right)^{5/2} \exp\left(-\frac{m}{T}\right) = 1. \qquad (12.23)$$

Somewhat surprisingly, an approximate solution to transcendental equations of this form is known and gives the result (Turner 1982)

$$\frac{m}{T_f} \simeq \ln Q + \frac{\ln(\ln Q)}{\frac{2}{5} - \ln Q} \; ; \; Q = \sqrt{\frac{90 n_a^2}{8\pi^3 g} \frac{M_P}{m}}. \qquad (12.24)$$

In terms of the freeze out temperature, the present day value of the monopole to photon ratio can therefore be estimated as

$$\left.\frac{n_m}{n_\gamma}\right|_{now} \sim \frac{g(T_{now})}{g(T_f)} \left(\frac{m}{T_f}\right)^{3/2} \exp(-\frac{m}{T_f}). \qquad (12.25)$$

Here the prefactor $g(T_{now})/g(T_f)$ accounts for the overall suppression of the monopole equilibrium number due to the decrease in the number of available effectively massless species. The result (12.25) shows the magnitude of the monopole problem. For typical values of the parameters (ignoring the species factors g for the present): the monopole mass $m \sim 10^{-3} M_P$ and $T_f \sim 10^{-1} m$, one has

$$\left.\frac{n_m}{n_\gamma}\right|_{now} \sim 10^{-9}. \qquad (12.26)$$

This is of the same order of magnitude as the present day value of the baryon to photon ratio. Since $m_{monopole} \sim 10^{16} m_{baryon}$ and there is expected to be roughly one monopole per baryon, the density parameter should be $\Omega_{now} \sim 10^{16}$, which is convincingly in disagreement with observation!

The best solution to the monopole problem, as to so many others, comes from cosmological inflation – the form of the solution was first described by Guth (1981). Remember that inflation, by expanding the universe exponentially by a factor $\exp(N)$, decreases the number densities of pre-existing particles by a factor $\exp(-3N)$. If the monopole number density freezes out *before* the start of inflation, $T_{freeze} \leq T_{inflate}$, then the present day number density of magnetic monopoles will be reduced by the inflation to

$$\left.\frac{n_m}{n_\gamma}\right|_{now} \sim \exp(-3N) \frac{g(T_{now})}{g(T_f)} \left(\frac{m}{T_f}\right)^{3/2} \exp(-\frac{m}{T_f}). \qquad (12.27)$$

For the inflationary models already examined above, N can easily be greater than 100 or 1000, so that, rather than dominating the energy density of our present universe, there can be less than a single magnetic monopole per horizon volume: $n_m/H^3 \ll 1$. Inflation solves the horizon problem in spectacular fashion. We must still approach the solution with some caution, however. Some small degree of tuning is necessary in order that the monopole problem does not resurface after inflation. Recall that the universe is reheated to a temperature, T_{reheat}, at the end of inflation: if $T_{reheat} \geq T_{freeze}$, then magnetic monopoles will be produced again after inflation, and the present day monopole to photon ratio will once more be given by (12.25). So in order for inflation to solve the monopole problem completely, the inflationary model must satisfy the main constraint that $T_{reheat} \ll T_{freeze}$.

12.3 Cosmological Formation of Boson Stars

Boson stars are one of the most interesting forms of exotic matter, particularly because they are possible dark matter candidates which would be very differently distributed at the present from other dark matter candidates.

In contrast with normal stars, boson stars are gravitationally bound macroscopic quantum states of scalar bosons. They are similar in many respects to neutron stars, differing in that their pressure support derives from the uncertainty relation rather than the exclusion principle. The possibility that boson stars may exist was demonstrated theoretically by Kaup (1968) and independently by Ruffini and Bonazzola (1969). Important advances in understanding the structure of boson stars were made by Colpi, Shapiro and Wasserman (1986). The cosmological formation of boson stars and their nature as dark matter candidates was first described by Madsen and Liddle (1990). Boson star structure and formation is extensively reviewed by Liddle and Madsen (1992).

Boson star structure can be well understood from simple dimensional arguments. The natural scale radius, r, of a gravitationally bound object of mass M is the size at which its gravitational energy becomes strong: $GM/r \sim 1$. Here the gravitational constant $G = 1/M_P^2$, where M_P is the Planck mass, and M is the total mass of the bound state. Now think of an individual free boson with mass, m. This has momentum $\sim m$ and thus a Compton wavelength, $r \sim 1/m$. Since bosons can occupy the same quantum state, a large number, n, of these particles can be accumulated in the same position, forming a state of radius r and total mass $M = nm$. This state is called a boson star. Since the maximum mass beyond which the star will collapse into a black hole is given by $GM/r \sim 1$, we can see that the maximum stable boson star mass is $M \sim M_P^2/m$. In this state, the energy density of the star is $\rho \sim M/r^3 \sim m^2 M_P^2$. Also, since the particles are bound, their kinetic energy is negligible, and the energy density can be written $\rho \sim V(\phi) \sim m^2 \phi^2$, so that in the bound state the scalar field must have the maximum possible value $\phi \sim M_P$.

Including the effects of a self-interaction term, $\lambda \phi^4$, in the scalar potential, $V(\phi)$, has a significant effect on the structure of the resulting boson star (Colpi, Shapiro and Wasserman, 1986). Consider the case with $\lambda \ll 1$

so that the effective mass of the bosons stays close to m and the equilibrium value of the field remains unchanged from the above calculation: $\phi \sim M_P$. The relative importance of the self-interaction is then measured by the dimensionless ratio $\lambda \phi^4 / m^2 \phi^2$. In the equilibrium bound state, this is $\Lambda \equiv \lambda M_P^2 / m^2$, showing that the interaction can dominate the potential even when λ is very small, provided only that $\lambda > m^2/M_P^2$. Thinking about this physically, this means that the self-interaction term introduces a new scale into the structure of the boson star equilibrium state: in addition to the Compton wavelength $r \sim 1/m$ of the individual particles, there is now the *coherence length* over which the interaction operates. Dimensionally, since $\Lambda = \lambda r^2 / r_p^2$ is a ratio of areas, the coherence length must be $R \sim \sqrt{\Lambda} r$. The bound state with these physical dimensions is then altered to have $GM/R \sim 1$, and this leads by the same reasoning used above to an estimate of the maximum mass of a boson star made from interacting particles: $M \sim \sqrt{\lambda} M_P^3/m^2$. The corresponding equilibrium value of the scalar boson field in such a state will be $\phi \sim \lambda^{-1/4} m$. It is clearly possible for the mass of a boson star made of interacting particles to be much larger than is possible with free (non-interacting) bosons. The physical interpretation of this cosmologically important fact is that the boson self-interaction (which is repulsive) acts like a pressure force, helping to hold up the star under its own weight.

The obvious cosmological question that arises from establishing the possible existence of boson stars is whether boson condensates could have formed in the early universe with the right sort of masses to become boson stars, rather than the alternative of forming primordial black holes (Khlopov, Malomed and Zeldovich, 1985 – see also section 12.1). The latter would happen if the typical mass of bosons condensing to each site exceeded M_P^2/m in the case that there was no self-interaction.

Early on, when the temperature $T > m$, the scalar field will remain spatially homogeneous on scales within the horizon size of $1/H$, since any perturbations will free stream away (Peebles 1971). It will also remain in thermal equilibrium, and $\rho_\phi \sim \rho_{rad} \sim T^4$ and the field's average value $\phi \sim T^2/m$. However, when the temperature drops below m, the boson field decouples from the radiation, and its amplitude freezes out at $\phi = a \sim m$. Perturbations in the field can then begin to grow if they are above the Jeans scale. Knowledge of the Jeans scale for a cold scalar field in the early universe requires an estimate of the pressure exerted by the field. Since the particles are cold, and thus stationary, one would expect them not to exert any pressure. However, due to the uncertainty relation, the bosons maintain a residual motion and so exert a small pressure which, dimensionally, can be estimated as $p \sim a\rho/M_P$. The Jeans wavenumber can then be calculated from the Jeans mass $M_J \sim M_P^3 P^{3/2}/\rho^2$, since $M_J \sim \rho l_J^3 \sim \rho k_J^3$. We find that $k_J^2 \sim m^2 a/M_P$. This is far smaller than the horizon scale at decoupling.

The boson star mass $M \sim M_P^2/m$ corresponds to a wavenumber $k_b^3 \sim (a^2/M_P^2)m^3$.

The mean growth rate, ω, of the perturbations is well estimated by the usual Newtonian value

$$\omega^2 \sim 4\pi G\rho \sim 4\pi \frac{m^2 a^2}{M_P^2} \tag{12.28}$$

for all wavelengths much smaller than the horizon size and much larger than the Jeans scale. Over this range, the perturbation growth rate will be mainly independent of scale. The corresponding collapse timescale is

$$t_c \sim \frac{1}{\sqrt{4\pi}} \frac{M_P^2}{ma} t_P, \tag{12.29}$$

where t_P is the Planck time. Note that at decoupling, $t_c \sim 1/H$, so we see that the approximation made is valid, since most of the collapse of a fluctuation occurs well within the horizon. On scales comparable with the horizon, however, the universe's expansion will logarithmically damp the exponential growth of perturbations. Any growth of the fluctuation on the largest scales is negligible compared to the exponential growth it experiences once it is well inside the horizon.

There is one problem with the formation process as it stands: at decoupling, $T \sim m$, so that the mass inside the horizon at decoupling, $M_d \sim \rho_\phi/H^3 \sim M_P^3/m^2$, is much larger than the boson star mass. Since the growth rate, ω, is insensitive to scale, most of the bosons inside the horizon will be swept up into *black holes* of mass $\sim M_d$ rather than boson stars. The problem is made even more serious by the fact that the growth of perturbations on scales well within the horizon is more likely to be damped by thermodynamical processes. Generally, horizon scale fluctuations (which are the smallest scales to survive free-streaming of the bosons) will have the largest amplitude. The horizon scale fluctuations will therefore be the dominant collapse scale.

The situation is more interesting for interacting bosons, since they form stable stars of mass $M \sim \sqrt{\lambda} M_d$. Models with $\lambda \sim 1$ then predict that essentially all the bosons will be swept up into boson stars by the gravitational collapse process. It is important to understand that even though the introduction of a nonzero λ completely alters the structure of the equilibrium bound states, it cannot have much effect on the growth rate of the linear perturbations on scales much larger than the Jeans length. The Jeans length itself will be increased due to the effective hydrodynamic pressure caused by the self-interaction, but this is unimportant relative to what is happening on scales corresponding to the horizon scale at decoupling.

The scenario presented so far is easily modified to take account of the possible existence of antibosons – the antiparticles corresponding to the scalar bosons. The breaking of particle–antiparticle symmetry at high temperature will result in this case in a boson–antiboson asymmetry of the same magnitude as the baryon–antibaryon asymmetry:

$$\frac{N_\phi}{n_\gamma} \sim \epsilon = \frac{n_B}{n_\gamma}, \tag{12.30}$$

where $N_\phi = n_\phi - n_{\bar\phi}$ is the net boson number. The net mass of bosons inside the horizon at decoupling is then reduced by a factor of ϵ to $M'_d \sim \epsilon M_d$ (assuming perfect annihilation). Correspondingly, the collapse timescale is increased to

$$t'_c \sim \frac{1}{\sqrt{4\epsilon\pi}} \frac{M_P^2}{ma} t_P, \tag{12.31}$$

The conclusion to be drawn from this discussion is that in physical models with $\lambda \geq \epsilon^2$, most of the net boson mass will form into boson stars. This conclusion holds regardless of the actual mass, m, of the individual bosons, although the collapse timescale is longer for lighter bosons. The number density of boson stars is also easily estimated: the typical hydrogen-burning star has a mass of about M_P^3/m_{proton}^2, so the ratio between the number of fermionic (hydrogen-burning) and boson stars is $N_{bs}/N_{fs} \sim m^3/\sqrt{\lambda} m_{proton}^3$. Notice the sensitivity of the number of boson stars formed to the boson mass. Also, this number only gives the mean cosmological density of boson stars. At this late stage of cosmic evolution, boson stars will probably be concentrated in the vicinity of visible galaxies. Table 12.1 displays a range of illustrative possibilities for the possible cosmological content of boson stars.

To summarize, then, a significant component of nonbaryonic dark matter may exist in the form of boson stars. This result has important implications for the distribution of dark matter. In particular, dark matter residing in boson stars will induce greater clustering power on short scales than other models in which the dark matter is more smoothly distributed.

12.4 Review

Black holes are objects whose radius is within their Schwarzschild radius (that radius at which the escape velocity at the surface of the object is equal to the speed of light), so that even light cannot escape from their surface. Because the horizon of the black hole provides a boundary for external radiation modes, black holes can shrink by emitting radiation from their horizon. This radiation is found to be thermal. The temperature of the radiation is inversely proportional to the mass of the black hole. The power radiated and the lifetime over which the black hole evaporates are easily calculated. The production of very small mass black holes in the early universe is ruled out by the smoothness of the cosmic background radiation.

Magnetic monopoles are predicted by many theories of elementary

Boson Star Properties

m	λ	ϵ	$M/M_\odot$	$M_d/M_\odot$	$Z(t_c)$
1 GeV	10^{-4}	10^{-9}	10^{-2}	10^{-9}	10^{13}
5 GeV	0	1	2×10^{-20}	4×10^{-2}	5×10^{13}
5 GeV	10^{-4}	1	4×10^{-4}	4×10^{-2}	5×10^{13}
5 GeV	10^{-4}	10^{-9}	4×10^{-4}	4×10^{-11}	5×10^{13}
5 GeV	1	10^{-9}	4×10^{-2}	4×10^{-11}	5×10^{13}
100 GeV	10^{-4}	10^{-9}	10^{-6}	10^{-13}	10^{15}

Table 12.1 *Selected model boson star parameters, showing the maximum boson star mass, M, and the horizon mass at decoupling, M_d, measured in units of the solar mass $M_\odot$. Boson stars may only form if the former exceeds the latter, $M > M_d$. Here, λ is the boson self-coupling, and ϵ is the boson–antiboson asymmetry parameter. The final column gives the redshift at the start of the collapse. Note that in the case of vanishing self-coupling the maximum value of the boson star mass is dramatically increased relative to the case where there is no self-coupling. (After Madsen and Liddle 1990, Liddle and Madsen 1992.)*

particles. They would have been produced in the early universe in large numbers through being in thermal equilibrium at temperatures greater than their mass. Naively, there would be about one monopole per baryon at the present time. Since magnetic monopoles are expected to be superheavy, with masses only a few orders of magnitude below the Planck mass, this means that the current value of the density parameter should be many orders of magnitude greater than unity. The fact that this prediction is clearly in contradiction with observation is known as the monopole problem. Detailed calculations of monopole production and annihilation show that the monopole problem remains severe across an enormous range of parameters. Fortunately, inflation solves the monopole problem by diluting their number density by the exponential of a large number, provided that the reheating temperature after inflation is well below the monopole mass.

Boson stars are compact gravitationally bound objects consisting of scalar bosons. Since they are cold, they are supported by the Heisenberg uncertainty principle. Free bosons can form only into relatively low mass stable objects, and estimates of the boson Jeans mass show that the stable mass of free bosons is much smaller than the mass that will be expected to collapse onto a fluctuation. Self-interacting bosons, on the other hand, can form stable objects with masses comparable to their horizon scale mass at the time of collapse. Because of their large masses, these boson stars will have extremely low number densities in the present universe. If the dark matter consists of boson stars, therefore, it will be almost entirely undetectable.

12.5 Exercises

Exercise 12.1 *Draw up a table showing the* radius, density, temperature *and* lifetime *of black holes having mass equal to that of:*

1. *a proton;*
2. *a human;*
3. *the earth;*
4. *the sun;*
5. *the Milky Way.*

Exercise 12.2 *Solve numerically the equations (12.21,12.22) for the magnetic monopole number density. Assume the number of annihilation channels, $n_a \sim 100$, the number of effectively massless particle species, $g \sim 100$, and the monopole mass, $m \sim 10^{-3} M_P$ (these are typical values for the simplest unified theories – see Kolb and Turner, 1990). Explore the parameter space in the neighbourhood of the values given here. Graph the solutions and locate the parameters for which the lowest monopole to photon ratios are achieved. Are these low enough to solve the monopole problem? Compare the numerical results with the estimates given in section 12.2 above.*

Exercise 12.3 *By deriving the perturbation equations for a scalar field in a homogeneous Newtonian background, show that the Jeans wavenumber for perturbations of a cold scalar field is given by*

$$k_J^2 = 4\sqrt{\pi}\frac{m^2 a}{M_P}$$

where a is the frozen value of the field amplitude: $\phi = a$.

12.6 References

Birrell, N.D. and Davies, P.C.W. *Quantum Fields in Curved Space* (Cambridge University Press, 1982).

Colpi, M., Shapiro, S.L. and Wasserman, I. *Phys. Rev. Lett.* **57**, 2485 (1986).

Guth, A.H. *Phys. Rev. D* **23**, 347 (1981).

Kaup, D.J. *Phys. Rev.* **172**, 1331 (1968).

Khlopov, M.Y., Malomed, B.A. and Zeldovich, Y.B. *Mon. Not. Roy. Astron. Soc.* **215**, 575 (1985).

Kolb, E.W. and Turner, M.S. *The Early Universe* (Addison Wesley, New York, 1990).

Liddle, A.R. and Madsen, M.S. *Int. J. Mod. Phys. D* **1**, 101 (1992).

Madsen, M.S. and Liddle, A.R. *Phys. Lett. B* **251**, 507 (1990).

Mandl, F. *Statistical Physics, 2ed.* (Wiley, Chichester, 1988).

McCrea, W.H. *Quart. J. Roy. Astron. Soc.* **27**, 137 (1986).

Peebles, P.J.E. *Physical Cosmology* (Princeton University Press, 1971).

Preskill, J.P. *Phys. Rev. Lett.* **43**, 1365 (1979).

Ruffini, R. and Bonazzola, S. *Phys. Rev.* **187**, 1767 (1969).

Turner, M.S. *Phys. Lett. B* **115**, 95 (1982).

13

Survey of Cosmological Theory

This book has been a journey through the history of the universe. Along the way, we have taken in many of the most interesting sights that are known. The items that we have seen and examined in close-up have helped us to build a picture of the universe: it was the purpose of this book to share that picture. Although many of the details of the picture may yet need improvement, we believe that some of the essentials are correct and these provide a framework within which to vary the models we have developed in the interest of developing ever better models.

13.1 What Have We Learned?

The universe was once hot and dense: driven by the energy of its own gravity, it expanded away from this initially hot dense state and cooled into the universe that we now know. Along the way, a small baryonic asymmetry was frozen out and left the relic matter from which all the luminous structures now visible eventually formed. At a much lower temperature, the neutrinos then decoupled from all other particles and have expanded freely to the present without very much contact with other matter. The radiation background was then heated slightly by the annihilation of electrons with positrons. The neutron to proton ratio decreased slowly due to the decay of free neutrons until the thermal energy of the baryons became low enough that they could join together into deuterium nuclei. These deuterium nuclei then formed mainly into helium nuclei, but so many of the neutrons had decayed that there were only enough neutrons to make about one-quarter of the mass of baryons into helium. Because the energy density of matter fell off more slowly than that of the radiation, the expansion of the universe eventually became dominated by the matter. When the temperature dropped below the ionization energy of hydrogen, the electrons and nuclei formed into atomic hydrogen and helium, with the result that the photon mean free path became unbounded and the universe became transparent.

This rendered Silk damping ineffective in preventing the growth of baryonic fluctuations, and these began to grow on all mass scales larger than the baryonic Jeans mass.

Meanwhile, fluctuations in any cold dark matter present had been stagnating on all mass scales smaller than the kinematic damping mass since the cold dark matter had dropped out of thermal equilibrium with the radiation. Once the radiation energy density ceased to dominate the expansion, though, the cold dark matter fluctuations on all scales resumed their growth. Those scales that had been beyond the horizon throughout the radiation dominated phase had grown much larger, so that the smallest structures to form correspond to those scales just above the kinematic damping mass.

Hot dark matter fluctuations, however, were damped by neutrino free streaming on all scales below the free streaming scale, and so none of them could grow until the horizon became larger than the neutrino free streaming length, whereupon the largest scales began to grow most rapidly. The cold dark matter fluctuations grew and clustered into larger fluctuations as these also grew, while the hot dark matter fluctuations flattened out into enormous pancakes, sweeping up and compressing the baryonic matter into the pancakes and triggering galaxy formation. The remaining baryons either grew their own fluctuations quite rapidly once their temperature dropped rapidly to zero after recombination, or fell into cold dark matter fluctuations, losing their kinetic energy through collisions and forming galaxies and clusters at the centre of cold dark matter structures. Once galaxies, quasars, and radio sources had formed these evolved slowly into the kinds of structures now observed in our universe.

The cosmological density parameter, which initially was relatively close to unity, continued to diverge away from unity as the expansion gradually slowed down, but remained within a few orders of magnitude of unity up to the present time.

Looking further back than the hot dense phase from which the universe expanded, the physical model which best fits our understanding of particle physics to the observed properties of the universe is the inflationary model. Inflationary models typically explain the magnitude of the initial fluctuations from which the observed structures grew as being given by the mass of the scalar inflaton field, and automatically generate a scale free spectrum of fluctuations because they create a period during which the universe is changing extremely slowly. They also smooth out any initially existing statistical fluctuations by their enormous exponential expansion. Such an exponential expansion also provides the solution to the horizon problem by simply inflating the horizon to such a size that our present observed universe corresponds to a tiny causally connected region of the universe from before the start of the inflationary period. During inflation, the universe is accelerating, so that the density parameter is also driven rapidly towards unity, solving the flatness problem. Any cosmologically undesirable objects, such as magnetic monopoles, which may exist before the start of inflation have their number densities diluted by the exponential inflationary factor, solving the monopole problem as well. Although inflation is not yet well understood in any great detail, the model of the inflationary universe generates the solutions to many of the cosmological problems, produces the kind of fluctuations required to form structures in reasonable agreement with those observed in the present day universe, and is based on a simple physical model with only one free parameter, the mass of the scalar inflaton.

13.2 The Present State of Cosmological Knowledge

Cosmological knowledge extends beyond what has been described in this book – hardly surprising since thus far we have studiously avoided discussions that involved general relativity and quantum field theory. For example, there are immensely sophisticated theories of relativistic cosmology (studying the universe using general relativity), theories of quantum fields in the early universe, and theories of structure formation based largely on enormous numerical simulations of the gravitational clustering of individual particles.

However, it would be fair to say that the material presented in this book represents nearly all of the core material of theoretical cosmology. Some of the details of theories described here may alter rapidly as time goes by and research advances, but the essentials will probably not be changed. It is, of course, precisely the essentials that this book has attempted to convey. Little space has been given to any of the details here: if the missing details were filled in, this book would have been closer to 10^3 pages in length than 10^2.

13.3 Problems Currently Facing Cosmology

In this book we have presented a unified approach to understanding how the universe has reached its present state. This involved presenting the successful and easily understandable problems, questions, and solutions in terms of physical and dynamical processes. However, there remain many questions about the universe in which we live, and about the models that we have described in this book. Some of the most puzzling are:

Inflation

Although the general characteristics of inflationary models are quite well known, particle physics has not yet produced a viable theoretical candidate for the scalar inflaton field, nor a framework within which its properties could be deduced. Inflationary models of the kind described in Chapter 9 must therefore be regarded as descriptive of the qualities expected to be possessed by inflationary models.

Dark matter

Despite the compelling evidence for the presence of some dark matter in the vicinity of galaxies and clusters as described in Chapter 11, there is no completely satisfactory model of structure formation based on dark matter. Simulations of hot dark matter models produce structures that are too tightly clumped to resemble our universe. Cold dark matter models produce more plausible looking structures, but require the assumption that baryonic structures are biased towards forming near peaks of the cold dark matter density distribution. No physical mechanism for biasing is yet understood. Furthermore, particle physics has not produced a viable candidate particle for the cold dark matter.

Density parameter

Observational measurements of the cosmological density parameter consistently result in estimates significantly below unity, despite the apparent inflationary prediction that it should be extremely close to unity. Understanding how this could happen requires closer examination of both the process of inflation and the evolution of the density parameter. Furthermore, detailed simulations of element synthesis in the early universe tend to restrict the baryonic contribution to the density parameter to an order of magnitude below unity.

Structure formation

Taken together, the previous pair of problems result in a major problem for our understanding of how structure could have formed. Structure has less time to form in a lower density universe, and purely baryonic structures are completely prevented from growing by Silk damping until after the end of the recombination epoch.

13.4 Reprise

Answers to these questions will require much research effort to be expended in the pursuit of knowledge about the composition and history of the universe in which we live. At the same time, it must be recognized that it is the magnitude of the questions and their answers that, above all, makes for the excitement and engagement that we experience in the study of our dynamic cosmos.

Appendix A

Physical Units and Constants

Throughout this book we have used physical units which are derived by combining the fundamental constants of physics so as to generate the *natural units* of measurement. In these units, the fundamental constants of thermodynamics (Boltzmann's constant, k), quantum theory (Planck's constant, $\hbar$), and relativity (the speed of light, c) are all chosen to be *dimensionless* and equal to unity:

$$\hbar = k = c = 1. \tag{A.1}$$

The fact that these constants have been chosen to be dimensionless means that we agree to measure:

- Distance in terms of the time it takes a light beam to travel that distance via the relation $x = ct$, so that a distance has the dimensions of time and the distance corresponding to 3×10^8 m can be written

$$t = \frac{3 \times 10^8}{c} \text{ s}$$

  which is approximately 1 second.

- Temperature in terms of the corresponding thermal energy of excitation per relativistic degree of freedom via the relation

$$E = kT$$

  so that temperature has the dimensions of energy and 1 degree Kelvin corresponds to approximately 10^{-4} eV.

- Momentum in terms of the corresponding Compton wavelength of a particle with that momentum via the relation $p\lambda = \hbar$, so that momentum has the dimensions of inverse length.

- Energy in terms of the equivalent rest mass with that energy via the relation

$$E = mc^2$$

so that energy has the dimensions of mass and 1 erg corresponds to approximately 10^{-21} g.

- Frequency in terms of the equivalent energy via the relation

$$E = \hbar\omega$$

so that, since $\omega = 2\pi/\lambda$, energy has the dimensions of inverse length.

Putting all this together, we can observe that energy, momentum, and temperature can all be expressed in terms of mass, since they have the same dimensions. Furthermore, length and time have the same dimensions, and can thus be expressed in terms of an inverse mass. The fundamental remaining dimension is that of mass, and we can choose an appropriate unit of mass in order to express all other physical quantities.

The only important physical relations that have not been displayed so far are those relating to gravity. It is natural to wonder whether we can reduce the fundamental constant of gravity (Newton's constant, G) to a dimensionless form as well. In fact, from the relation for the gravitational self-energy of an object in terms of its mass and radius

$$E \sim \frac{GM}{r^2}$$

we can see that $\sqrt{G}$ must have the same dimensions as length or equivalently, inverse mass. Choosing G to be dimensionless would therefore destroy the final remaining physical scale, so that this cannot be allowed if we want to be able to compare the predictions of physical theory with observations. The remaining scale can be chosen to be a mass, and this is normally taken in natural units to be the *Planck mass*,

$$M_P = \frac{1}{\sqrt{G}}$$

so called because it was first derived by Max Planck around the turn of the century.

The above discussion makes clear how to perform calculations using natural units. First, check that any relation is dimensionally correct – this part is easy when using natural units. Next, calculate the desired result, preferably as a dimensionless ratio times some power of the Planck mass. Then convert the result into the desired type of quantity by using the conversion factors given below for the Planck mass, Planck length, and Planck time.

The physical constants necessary for the purposes of this book are given (to sufficient accuracy for cosmological calculations) in the accompanying tables.

Planck Units

Planck mass	M_P	10^{19} GeV
Planck time	t_P	10^{-43} s
Planck length	l_P	10^{-33} cm

Table A.1 *The Planck units expressed in conventional measurement units of energy, length, and time.*

Energy Conversion Factors

Temperature	1 GeV	10^{13} K
Mass	1 GeV	10^{-24} g
Time	1 GeV^{-1}	10^{-24} s
Length	1 GeV^{-1}	10^{-14} cm

Table A.2 *The conversion factors used to relate the usual measures of temperature, mass, time, and length to the energy-based system of units.*

Fundamental Physical Constants

Planck constant	$\hbar$	1.1×10^{-27} cm^2.g.s^{-1}
Speed of light	c	3×10^{10} cm.s^{-1}
Boltzmann constant	k	1.4×10^{-16} erg.K^{-1}
Newton constant	G	6.7×10^{-8} cm^3.g^{-1}.s^{-1}
Fermi constant	G_F	1×10^{-5} GeV^{-2}
Thomson cross-section	σ_T	6.7×10^{-25} cm^2
Blackbody constant	$a = \frac{\pi^2 k^4}{15 c^3 \hbar^3}$	7.6×10^{-15} erg.cm^{-3}.K^{-4}
Electron mass	m_e	$\frac{1}{2}$ MeV
Proton mass	m_p	1 GeV
Neutron mass	m_n	1 GeV
Neutron–proton mass split	$m_n - m_p$	1.3 MeV

Table A.3 *The fundamental constants of physics which are necessary in the study of cosmology.*

Astronomical Constants

Solar mass	$M_\odot$	2×10^{33} g
	$M_\odot$	10^{57} GeV
Hubble time	$t_H = 1/H_{now}$	$3 \times 10^{17} \left(\frac{100 \text{km.s}^{-1}.\text{Mpc}^{-1}}{H_{now}} \right)$ s
Parsec	pc	3×10^{18} cm
	pc	3 light-year

Table A.4 *The astronomical constants which most frequently arise in calculations relating to cosmology.*

Appendix B

Selected Bibliography of Cosmology

B.1 Introductory References

The introductory references listed here should all be accessible to any interested reader, since they assume no prior knowledge of physics, and do not require either familiarity with, or use of, any mathematics. Since there are a large number of such works available, the following cannot be given as an exhaustive list. Rather, I have selected some of those which I have myself found interesting and enjoyable.

- John D. Barrow, *The Origin of the Universe* (Orion, Weidenfeld and Nicolson, London, 1994) ISBN 0-297-81497-4; (Basic Books, New York, 1994) ISBN 0-465-05-354-8.
  A general account at the popular level of most of the material covered in this book.

- Paul Davies, *The Cosmic Blueprint* (Heinemann, London, 1987) ISBN 0-04-440182-5.
  A smoothly written account of the progress of the universe from its earliest stages to the present.

- Heinz R. Pagels, *The Cosmic Code: Quantum Physics as the Language of Nature* (Michael Joseph, London, 1982) ISBN 0-7181-2217-8.
  Examines how the quantum nature of matter was deduced and how this knowledge aids our understanding of the way the universe appears now.

- Steven Weinberg, *The First Three Minutes* (André Deutsch, London, 1977) ISBN 0-00-654024-4.
  The definitive popular book on cosmology by a Nobel laureate physicist. Necessary reading for anyone interested in the subject. Clear and forceful prose, lucid explanations, and enviably simple mathematical agruments presented in an appendix. It has been reprinted a number of times under different imprints and in different revisions.

Frank Wilczek and Betsy Devine, *Longing for the Harmonies: Themes and Variations from Modern Physics* (W.W. Norton and Co, New York, 1987) ISBN 0-393-30596-1.
An unusual attitude to cosmology. This book dips into the cosmic questions and provides selected answers in a thought-provoking fashion.

B.2 Intermediate References

The following list contains some of the books which are most similar in spirit to the present work. They mainly assume that the reader has only a minimal knowledge of the physics involved in understanding the universe, but at least occasionally they make some mathematical demands on the reader.

John D. Barrow and Frank J. Tipler, *The Anthropic Cosmological Principle* (Oxford University Press, 1986) ISBN 0-19-851949-4.
This is an intellectually demanding book which will be most appreciated by the philosophically minded reader. The middle chapters contain a large amount of clear, intuitive derivations of important cosmological relations.

P.C.W. Davies, *The Accidental Universe* (Cambridge University Press, 1982) ISBN 0-521-242126.
A technical book from a consummately skilled popular author, this volume is written in Paul Davies' usual exciting style, with the mathematics used to enhance the flow of the argument.

I.D. Novikov, *Evolution of the Universe* (Cambridge University Press, 1983) ISBN 0-521-24129-4.
Similar to *The Accidental Universe* in content, but written in a more conventional style.

Michael Rowan-Robinson, *Cosmology* (Clarendon Press, Oxford, 1977) ISBN 0-19-851838-2.
Essential reading for anyone seriously interested in the universe because of the excellent and accessible descriptions of cosmological observations and how they are interpreted.

I.L. Rozental, *Big Bang Big Bounce: How Particles and Fields Drive Cosmic Evolution* (Springer-Verlag, Berlin, 1988) ISBN 3-540-17904-6.
Highly individual treatment of the anthropic principle as it applies to cosmology.

D.W. Sciama, *Modern Cosmology* (Cambridge University Press, 1971) ISBN 0-521-08069-X.
This book was written during one of the most intense periods of activity in cosmology this century. It is absolutely essential reading for the sense of excitement in the cosmological *zeitgeist* that it conveys. In addition,

it contains excellent syntheses of the observational data available at that time.

Joseph Silk, *The Big Bang* (W.H. Freeman, San Francisco, 1980) ISBN 0-7167-1084-6.
One of those books that one cannot read without wishing that one had written it oneself. It is calm, unhurried and thorough in its presentation of the subject. The text itself is accessible to anyone with a normal vocabulary, while the mathematical notes display the author's powerful insight into the hard questions by showing how simple the answers can be. Essential reading.

B.3 Advanced References

These references contain more detailed material than is presented here, and require correspondingly greater effort from the reader. At the same time, they are the definitive sources for most of the current understanding of cosmology, apart from the academic journals devoted to the subject.

Gerhard Börner, *The Early Universe: Facts and Fiction* (Springer-Verlag, Berlin, 1988) ISBN 3-540-16187-2.
Admirably thorough and even in its presentation of mainstream cosmology. The only textbook that contains an account of the relativistic theory of observational cosmology. Excellent treatment of structure formation that goes well beyond what is covered in this volume. Well worth owning for the cartoons alone. Especially useful for anyone wishing to do research in cosmology.

James Jeans, *Astronomy and Cosmogony* (Cambridge University Press, 1928) ISBN X-12-031953-4.
A classic text from that happier time when most of what was known about astronomy, astrophysics, and cosmology could be fitted between one set of covers. Clarity and elegance of style that have never dated even as the content has.

Edward W. Kolb and Michael S. Turner, *The Early Universe* (Addison-Wesley, New York, 1990) ISBN 0-201-11603-0.
Another essential book for anyone who wishes to do research in cosmology. Contains very detailed and highly technical accounts of nucleosynthesis, baryon genesis, inflation, and structure formation, performing comparative analyses of different models and mechanisms in each case. Much of the material covered is otherwise available only in the original literature. A valuable resource for anyone not intimidated by general relativity and quantum field theory on the same page.

Andrei Linde, *Particle Physics and Inflationary Cosmology* (Harwood, Chur, 1990) ISBN 3-7186-0489-2.

This book covers mainly the material named in the title, concentrating on the inflationary universe. Supremely clear explanations given by the principal architect of the chaotic inflation model.

P.J.E. Peebles, *Physical Cosmology* (Princeton University Press, 1971) ISBN 0-691-08137-9.
Full of calculations which will tax much of the physical knowledge possessed by the reader, this book adopts a rigorous approach of comparison between theory and observation. Some of the issues it touches are now well out of the mainstream of cosmology, but it remains essential – and highly educational – reading.

P.J.E. Peebles, *The Large Scale Structure of the Universe* (Princeton University Press, 1980) ISBN 0-691-08240-5.
More tightly structured than his previous book, this work focuses entirely on the growth of structure and its determination from direct observation. While these research areas have since advanced a long way, this book lays the groundwork for understanding the later research. Absolutely essential for anyone curious about large scale structure.

D.J. Raine, *The Isotropic Universe* (Adam Hilger, Bristol, 1981) ISBN 0-85274-370-X.
An excellent book, particularly clear and strong on observational aspects of cosmology and on the interpretation of observations within the framework of general relativity. Especially worth reading is the exposition of the mathematical theory of the standard cosmological models.*

B.4 Relativistic Cosmology

These works cover cosmology by starting from the viewpoint of general relativity, and then move on to the derivation of the field equations describing the universe. Because general relativity permits far more complicated model universes than the basic one studied in the present volume, these references contain many calculations and results additional to what is found here.

L.D. Landau and E.M. Lifshitz *The Classical Theory of Fields, 4ed.* (Pergamon, Oxford, 1975) ISBN 0-08-025072-6.
Taking an unashamedly physical approach to relativity, this book is more concerned with finding relations between quantities than exploring their abstract properties. The sections on relativistic cosmology are straightforward, if mathematically dense, and cover material hardly touched on in this work.

* The following gem is to be found in the preface: 'It is too much of a cliche to express the hope that this book will encourage a handful of its readers to further study and insight. In fact, the result I think I should really like is that the readers of this book should find it entertaining. For to be entertained is, I suppose, the only sane response to a Universe so scaffolded in misconception.'

Wolfgang Rindler *Essential Relativity: Special, General, and Cosmological, 2ed.* (Springer-Verlag, New York, 1977) ISBN 0-387-10090-3.
Easily one of the most readable physics book ever written, this book by the author who formalized the horizon concept steps the reader through a series of easy arguments into a clear understanding of the need for general relativity. The discussion of the isotropic and homogeneous cosmological models is a model of clarity.

Steven Weinberg, *Gravitation and Cosmology: Principles and Applications of the General Theory of Relativity* (Wiley, New York, 1972) ISBN 0-471-92567-5.
One of the modern classic texts. It contains a full treatment of most aspects of general relativity starting from a formal physical viewpoint. The sections on physical cosmology in particular are starkly clear. In common with many of its contemporaries, its coverage of the subject area has shrunk as cosmology has expanded its reach, but nearly all of the material contained in this large book remains of interest to the serious student and researcher.

Yakov B. Zeldovich and Igor D. Novikov, *Relativistic Astrophysics*;

Volume I: Stars and Relativity (University of Chicago Press, 1982) ISBN 0-226-97955-5.

Volume II: Structure and Evolution of the Universe (University of Chicago Press, 1983) ISBN 0-226-97957-1.

The authors of these volumes have contributed large areas to modern cosmology almost single handed. Their treatment is often disarmingly straightforward, and there is very little in the second volume that is not still of interest to researchers in the area. Newcomers to cosmology should expect to find the going hard until they develop their physical intuition to the same level as the authors.

B.5 Selected Journal Articles

The articles listed here are mainly papers describing original research, and as such they demand of the reader a considerable background in the subject. Most of them are included here for historical interest, since nearly all of the results they describe are presented elsewhere, in more recent reviews and texts.

M. Clutton-Brock, *Quart. J. Roy. Astron. Soc.* **34**, 411 (1993).
Contains clear and accessible descriptions of the cosmological problems and how they are solved by inflation.

A.A. Friedmann, *Z. Phys.* **10**, 377 (1922).

A.A. Friedmann, *Z. Phys.* **21**, 326 (1924).
The original papers in which the equations and solutions corresponding to an expanding universe were derived and studied.

C. Hayashi, *Prog. Theor. Phys.* **5**, 224 (1950).
A difficult paper to read, this shows how the nuclear reaction rates conspire to fix simple initial conditions for nucleosynthesis.

A.D. Linde, *Phys. Lett.* **129B**, 177 (1983).
The original paper describing the chaotic inflationary model.

M.S. Madsen and P. Coles, *Nucl. Phys. B* **298**, 701 (1988).
The first paper to show that chaotic inflation does not require special initial conditions.

W.H. McCrea and E.A. Milne, *Quart. J. Math. (Oxford)* **5**, 73 (1934).
The article in which it was first shown that Newtonian theory predicts the same equations for the evolution of the universe as those derived from relativistic theory.

E.A. Milne, *Quart. J. Math. (Oxford)* **5**, 64 (1934).
In this paper, kinematic arguments were used to explain Hubble's law.

W. Rindler, *Mon. Not. Roy. Astr. Soc.* **116**, 662 (1956).
The original formalization of the horizon concept, this is probably still the clearest introduction to the subject.

B.6 General Physics

These references are included because they cover aspects of the physics used in the present volume, such as the statistical physics and physics of particles.

I.R. Kenyon, *Elementary Particle Physics* (Chapman & Hall, London, 1988) ISBN 0-7102-1234-8.
This book does for particle physics what the present volume hopes to do for cosmology: simplifies the subject down to the essentials so as to give a sound technical grounding for a more thorough study. Neatly written and presented.

F. Mandl, *Statistical Physics, 2ed.* (Wiley, Chichester, 1988) ISBN 0-471-91533-5.
In this author's experience, the best book to read in order to understand thermal and statistical physics, and the only one ever to make any of the subject interesting.

Abraham Pais, *Inward Bound: Of Matter and Forces in the Physical World* (Oxford University Press, 1986) ISBN 0-19-851971-0.
By the author of a fine biography of Einstein, a challenging general overview of fundamental physics. There are very few people who could claim not to have anything to learn from this book.

D.H. Perkins, *Introduction to High Energy Physics, 2ed.* (Addison-Wesley, Reading, Mass., 1982) ISBN 0-201-05757-3. The classic (and essential) text for anyone who wants to understand what particle physics theories are, what particles they describe, how experiments are related to theory, and how to understand the importance of symmetry.

B.7 Complete List of Articles

This section contains a complete listing of all article references occurring within the text of this book.

Abell, G. *Astrophys. J. Suppl.* **3**, 211 (1958).

Alpher, R.A., Bethe, H.A. and Gamow, G. *Phys. Rev.* **73**, 803 (1948).

Barrow, J.D. *Quart. J. Roy. Astron. Soc.* **29**, 101 (1988).

Barrow, J.D. *Quart. J. Roy. Astron. Soc.* **30**, 163 (1989).

Blumenthal, G.R., Faber, S.M., Primack, J.R. and Rees, M.J. *Nature* **311**, 517 (1984).

Bonnor, W.B. *Mon. Not. Roy. Astron. Soc.* **117**, 104 (1957).

Clutton-Brock, M. *Quart. J. Roy. Astron. Soc.* **34**, 411 (1993).

Colpi, M., Shapiro, S.L. and Wasserman, I. *Phys. Rev. Lett.* **57**, 2485 (1986).

Davies, R.D. *Quart. J. Roy. Astron. Soc.* **29**, 443 (1988).

Dicke, R.H. and Peebles, P.J.E. in Hawking, S.W. and Israel, W. (eds) *General Relativity: An Einstein Centenary Survey* (Cambridge University Press, 1979).

Dicke, R.H., Peebles, P.J.E., Roll, P.G. and Wilkinson, D.T. *Astrophys. J.* **142**, 414 (1965).

Friedmann, A.A. *Z. Phys.* **10**, 377 (1922).

Friedmann, A.A. *Z. Phys.* **21**, 326 (1924).

Guth, A.H. *Phys. Rev. D* **23**, 347 (1981).

Guth, A.H. and Pi, S.-Y. *Phys. Rev. Lett.* **49**, 1110 (1982).

Harrison, E.R. *Phys. Rev. D* **1**, 2726 (1970).

Hawking, S.W. *Phys. Lett. B* **115**, 295 (1982).

Hayashi, C. *Prog. Theor. Phys.* **5**, 224 (1950).

Iben, I. and Renzini, A. *Phys. Rep.* **105**, 331 (1984).

Kaup, D.J. *Phys. Rev.* **172**, 1331 (1968).

Khlopov, M.Y., Malomed, B.A. and Zeldovich, Y.B. *Mon. Not. Roy. Astron. Soc.* **215**, 575 (1985).

Kolb, E.W. and Wolfram, S. *Nucl. Phys. B* **172**, 224 (1980).

Liddle, A.R. and Madsen, M.S. *Int. J. Mod. Phys. D* **1**, 101 (1992).

Lifshitz, E.M. and Khalatnikov, I.M. *Adv. Phys.* **12**, 185 (1963).

Linde, A.D. *Phys. Lett. B* **129**, 177 (1983).

Linde, A.D. *Phys. Lett. B* **162**, 281 (1985).

Linde, A.D. *Phys. Lett. B* **175**, 395 (1986).

Madsen, M.S. and Coles, P. *Nucl. Phys. B* **298**, 701 (1988).

Madsen, M.S. and Ellis, G.F.R. *Mon. Not. Roy. Astron. Soc.* **234**, 67 (1988).

Madsen, M.S. and Liddle, A.R. *Phys. Lett. B* **251**, 507 (1990).

Madsen, M.S., Mimoso, J.P., Butcher, J.A. and Ellis, G.F.R. *Phys. Rev. D* **46**, 1399 (1992).

Mather, J.C. *et al. Astrophys. J.* **354**, L37 (1990).

Mather, J.C. *et al. Astrophys. J.* **420**, 439 (1994).

McCrea, W.H. *Quart. J. Roy. Astron. Soc.* **27**, 137 (1986).

McCrea, W.H. and Milne, E.A. *Quart. J. Math. (Oxford)* **5**, 73 (1934).

Meszaros, P. *Astron. Astrophys.* **38**, 5 (1975).

Milne, E.A. *Quart. J. Math. (Oxford)* **5**, 64 (1934).

Peebles, P.J.E. *Astrophys. J.* **146**, 542 (1966).

Penzias, A.A. and Wilson, R.W. *Astrophys. J.* **142**, 419 (1965).

Preskill, J.P. *Phys. Rev. Lett.* **43**, 1365 (1979).

Rindler, W. *Mon. Not. Roy. Astr. Soc.* **116**, 662 (1956).

Rowan-Robinson, M., Walker, D. and Yahil, A. *Astrophys. J.* **301**, L1 (1986).

Ruffini, R. and Bonazzola, S. *Phys. Rev.* **187**, 1767 (1969).

Sachs, R.K. and Wolfe, A.M. *Astrophys. J.* **147**, 73 (1967).

Sakharov, A.D. *Zh. Eksp. Teor. Fiz. Pisma Red.* **5**, 32 (1967).

Shapley, H. and Ames, A. *Ann. Harvard College Obs.* **88**, 41 (1932).

Silk, J. *Astrophys J.* **151**, 459 (1968).

Silk, J. *Can J. Phys.* **64**, 147 (1986).

Songaila, A. *et al. Nature* **371**, 43 (1994).

Toussaint, D., Treiman, S.B., Wilczek, F. and Zee, A. *Phys. Rev. D* **19**, 1036 (1979).

Turner, M.S. *Phys. Lett. B* **115**, 95 (1982).

Vandenberg, D.A. *Astrophys. J. Suppl. Series* **51**, 24 (1983).

Weinberg, S. *Phys. Rev. Lett.* **42**, 850 (1979).

Zeldovich, Y.B. *Mon. Not. Roy. Astron. Soc.* **160**, 1P (1972).

B.8 Complete List of Books

This section contains a complete listing of all book references occurring within the text of this book.

Barrow, J.D. and Tipler, F.J. *The Anthropic Cosmological Principle* (Oxford University Press, 1986).

Bertotti, B., Balbinot, R., Bergia, S. and Messina, A. *Modern Cosmology in Retrospect* (Cambridge University Press, 1990).

Birrell, N.D. and Davies, P.C.W. *Quantum Fields in Curved Space* (Cambridge University Press, 1982).

Börner, G. *The Early Universe: Facts and Fiction* (Springer-Verlag, Berlin, 1988).

Darwin, C. *On the Origin of Species by Means of Natural Selection* (John Murray, London, 1859).

Gould, S.J. *Time's Arrow, Time's Cycle* (Harvard University Press, Cambridge, Massachussetts, 1987).

Hubble, E.P. *The Realm of the Nebulae* (Yale University Press, 1936).

Jeans, J. *Astronomy and Cosmogony*, (Cambridge University Press, 1928).

Kolb, E.W. and Turner, M.S. *The Early Universe* (Addison-Wesley, New York, 1990).

Mandl, F. *Statistical Physics, 2ed.* (Wiley, Chichester, 1988).

Peebles, P.J.E. *Physical Cosmology*, (Princeton University Press, 1971).

Raine, D.J. *The Isotropic Universe* (Adam Hilger, Bristol, 1981)

Rowan-Robinson, M. *Cosmology* (Clarendon Press, Oxford, 1977).

Sciama, D.W. *Modern Cosmology* (Cambridge Unversity Press, 1971).

Silk, J. *The Big Bang* (W.H. Freeman, San Francisco, 1980).

Stauffer, D. and Aharony, A. *Percolation Theory, 2ed.* (Taylor and Francis, London, 1992).

Weinberg, S. *Gravitation and Cosmology* (Wiley, New York, 1972).

Weinberg, S. *The First Three Minutes* (André Deutsch, London, 1977).

Zeldovich, Y.B. and Novikov, I.D. *Structure and Evolution of the Universe* (University of Chicago Press, 1983).

Index

Abell, G., 18, 137
Acoustic wave, 92
Adiabatic expansion, 50
Adiabatic index, 40, 45, 49
 effective, 43, 74
Aharony, A., 100, 108, 139
Alpher, R.A., 18, 137
Ames, A., 18, 138
Andromeda galaxy, 6, 98, 99
Antibaryons, 68, 70, 119
Antibosons, 118, 119
Antimatter, 10, 11, 67–70
 in cosmic rays, 11
 matter–antimatter asymmetry, 12
Antineutron, 60
Antiparticle, 10, 11, 69
Antiproton, 60
Asymmetry, 67–70, 123

Balbinot, R., 18, 139
Barrow, J.D., 8, 15, 18, 79, 84, 95, 96, 131, 132, 139
Baryons, 57, 60, 67–70, 80, 94, 99, 101, 103, 104, 106, 107, 112, 115, 120, 123, 124
Bergia, S., 18, 139
Bertotti, B., 15, 17, 18, 139
Beryllium, 62
Beta decay, 62, 63
Bethe, H.A., 18, 137
Big bang, 50, 52, 53, 57, 61, 76, 80
Binding energy
 gravitational, 50
 nuclear, 59
 of deuterium, 63
Birrell, N.D., 121, 139
Black holes, 2, 90, 109–112, 116–119, 121
 primordial, 110
Blackbody constant, 52
Blackbody radiation, 52
Blumenthal, G.R., 104, 108, 137
Boltzmann constant, 53, 91, 127
Boltzmann factor, 57, 62
Bonazzola, S., 116, 122, 138
Bonnor, W.B., 96, 137
Börner, G., 11, 13, 17, 18, 133, 139
Boron, 62
Boson condensates, 117
Boson stars, 2, 109, 116–121
Butcher, J.A., 47, 138

Causal connection, 35, 36, 40, 72, 124
Causality, 16, 17
 problem, 16
Chaotic inflation, 75
Classical era, 83
Clusters, 6, 102, 107, 124, 126
 galaxy clusters, 14, 85, 93, 97–99, 102, 108
 globular clusters, 9, 10, 35, 108
  age of, 9
 superclusters, 6, 13, 93, 103, 108
Clutton-Brock, M., 16, 18, 79, 84, 135, 137

Index 141

Coherence length, 117
Coles, P., 76, 84, 136, 138
Collisional damping, 104
Colpi, M., 116, 121, 137
Comoving coordinates, 24, 33, 36, 85, 87
Compton wavelength, 116, 117, 127
Copernican principle, *see* Cosmological principle
Cosmic background radiation, 12–13, 16, 17, 52, 59, 78, 86, 89, 93–96, 119
    isotropy, 13
    spectrum, 12
    temperature, 12, 15
    temperature distribution, 12
Cosmic neutrino background, 56
Cosmic rays, 11, 67
Cosmological constant, 75
Cosmological density parameter, *see* Density parameter
Cosmological principle, 21, 77, 95
Cosmological problems, 71–72
Cosmology, xii, 67, 68, 72, 133
Critical density, 13, 17, 112

Dark matter, 2, 14, 94, 97–108, 112, 116, 119, 121, 126
    cold, 97, 104–107, 124, 126
        cutoff scale, 105, 107
    evidence for, 97–99
    hot, 97, 101–104, 107, 124, 126
    problem, 97
Darwin, C., 8, 18, 139
Davies, P.C.W., 121, 131, 132, 139
Davies, R.D., 12, 18, 137
De Sitter, W., 1
Deceleration parameter, 39, 41
Decoupling, 56–58, 100, 108
    epoch, 57
    of boson field, 117, 118
    temperature, 57
Degrees of freedom, 55, 127
Density contrast, 87
Density fluctuations, 78, 80, 81, 85–96
Density parameter, 2, 13, 38–47, 58, 71, 77, 89, 96, 97, 112, 120, 124, 126
Detailed balance, 113

Deuterium, 62–65, 123
Devine, B., 132
Dicke, R.H., 17, 18, 52, 58, 137
Dirac delta function, 54, 61
Doppler
    effect, 36
    formula, 35
    shift, 6, 98
Dust, 38, 49

Einstein, A., 1
Electromagnetic charge, 111
Electromagnetic radiation, 50
Electron–positron
    annihilation, 56, 57, 123
    gas, 57
    pairs, 55, 111
Element abundances, 10, 60
    cosmological, 63
Ellis, G.F.R., 43, 47, 138
Energy conservation, 30
Entropy, 23, 50–52, 55
Equation of continuity, 86
Equation of state, 23, 30, 40, 47, 49
Euler equation, 88
Euler–Lagrange equation, 73
Evolution
    dynamical, 2
Expansion, 2
Expansion equations
    derivation, 21–25
    solutions, 26–30

Faber, S.M., 108, 137
Fermi factor, 54, 61
First law of thermodynamics, 23, 50
Flatness, 39
Flatness problem, 39–41, 45, 71, 77, 97, 125
Fourier coefficients, 91
Fourier component, 92
Fourier expansion, 91
Fourier sum, 91
Freeze-out temperature
    of weak interactions, 55
Friction dominated regime, 82, 83
Friedmann, A.A., 1, 25, 31, 135–137

Galactic rotation curves, 97–99
Galaxy formation, 12
Gamma radiation, 67, 111
Gamow, G., 18, 137
Gauge boson, 70
Gauge symmetry, 69
Gaussian curvature, 45
Geological age of the earth, 8
Gould, S.J., 8, 18, 139
Gravitational waves, 82
Guth, A.H., 78, 84, 115, 121, 137

Harrison, E.R., 79, 84, 137
Hawking, S.W., 18, 78, 84, 137
Hayashi, C., 61, 65, 136, 137
Heisenberg uncertainty principle, 120
Helium, 10, 123
    abundance, 10, 17
    mass fraction, 59, 62–65, 67
Horizon, 2, 33, 36–38, 42, 46, 81, 100, 103, 106, 118–119, 121, 124
    black hole, 110, 111, 119
    event horizon, 37, 38, 47
    existence of, 46
    horizon problem, 17, 40, 71, 72, 77, 115, 124
    horizon volume, 115
    particle horizon, 37, 38, 42, 47, 72
    physical horizon, 72
    scale at decoupling, 118
    size at decoupling, 94
Hubble parameter, 9, 26, 36, 40, 41
Hubble's constant, 7, 8, 25
Hubble's law, 14, 17, 21, 25–26, 86
Hubble, E.P., 7, 8, 14, 18, 38, 139
Hydrodynamical equations, 86
Hydrogen, 10

Iben, I., 9, 18, 137
Inflation, 2, 45, 71–84, 86, 97, 105, 109, 110, 115, 116, 120, 124–126
    chaotic, 75–77
    eternal chaotic, 81–83
Inflaton, 71, 76, 78, 79, 81–84
    mass of, 80
Invisible material, *see* Dark matter
Ionization energy

of hydrogen, 57, 58, 123
Ionizing photons, 57
Iron ball theorem, 22
Isothermal gas law, 91
Israel, W., 18, 137

Jeans length, 93, 106, 118
Jeans mass, 81, 86, 93, 104, 106, 108, 117, 120, 124
Jeans scale, 89, 96, 101, 102, 117, 118
Jeans wavenumber, 117, 121
Jeans, J., 88, 92, 96, 133, 139

Kaup, D.J., 116, 121, 137
Kenyon, I.R., 136
Khalatnikov, I.M., 96, 138
Khlopov, M.Y., 117, 121, 137
Kolb, E.W., 17, 18, 69, 70, 121, 122, 133, 137, 139

Lagrangian function, 73, 76
Landau, L.D., 134
Large scale structure, 2, 6, 15, 17, 42, 56, 71, 76, 85, 89, 95, 97, 104, 107, 126
Liddle, A.R., 116, 120, 122, 138
Lifshitz, E.M., 96, 134, 138
Light signal, 33–34
    emission of, 34
    reception of, 34
Linde, A.D., 76, 78, 81, 84, 133, 136
Lithium, 62
Local Group, 98

M31, *see* Andromeda galaxy
Madsen, M.S., 43, 47, 76, 84, 116, 120, 122, 136, 138
Magnetic monopoles, 2, 109, 112–116, 119–121, 125
    annihilation, 112, 113
    monopole problem, 112, 115, 116, 120, 121
Malomed, B.A., 117, 121, 137
Mandl, F., 122, 136, 139
Mass to luminosity ratio, 13
Mather, J.C., 18, 138

Index

Matter era, 58
McCrea, W.H., 25, 31, 110, 122, 136, 138
Mean value theorem, 34
Messina, A., 18, 139
Meszaros, P., 105, 108, 138
Metaphysics, 45
Milky Way, 6, 7, 9, 98, 121
Milne, E.A., 25, 31, 136, 138
Mimoso, J.P., 47, 138
Monopoles, see Magnetic monopoles

Natural units, 52, 127
Neutrino decoupling, 54, 55
Neutrino free streaming, 102, 103, 107, 108, 124
Neutrino haloes, 102
Neutrino temperature, 54, 58, 108
Neutron lifetime, 64, 65
Neutron stars, 116
Neutron to proton ratio, 59–65, 123
Newton's constant, 86, 128
Newton, I., 22
Novikov, I.D., 10, 19, 132, 135, 139
Nuclear fusion, 10
Nuclear reactions, 53
Nucleosynthesis era, 60, 61

Observations, 1, 5, 17, 63–65, 97, 107
observe, 5–19
Occam's razor, 21
Ω, see Density parameter

Pagels, H.R., 131
Pais, A., 136
Pancake model, 104, 124
Pauli exclusion principle, 102
Peebles, P.J.E., 10, 14, 17, 18, 35, 38, 58, 65, 93, 96, 117, 122, 134, 137–139
Penzias, A.A., 18, 52, 58, 138
Perkins, D.H., 137
Perturbation problem, 71
Perturbations, see Density fluctuations
Phase space, 54, 61, 102

Photon mean free path, 57, 123
Pi, S.-Y., 78, 84, 137
Pions, 59
Planck length, 81, 128
Planck limit, 77
Planck mass, 74, 77, 83, 112, 116, 120, 128
Planck time, 83, 118, 128
Planck units, 129
Planck's constant, 127
Planck, M., 128
Plasma
  electron–proton, 56
Poisson equation, 79
Preskill, J.P., 112, 114, 122, 138
Primack, J.R., 108, 137

Quantum corrections, 77
Quantum fluctuations, 78
Quasars, 15–17, 35, 124

Radiation era, 58, 76, 124
Radio sources, 15, 17, 124
  counts of, 16
Raine, D.J., 14, 18, 98, 108, 134, 139
Random walk, 100
Recombination, 56–57, 101, 106, 124, 126
Redshift, 7, 33–36, 38, 42, 45, 53, 58, 89
  cosmological, 2, 41
  relation, 35
  velocity, 36, 38
Rees, M.J., 108, 137
Relativity
  general, 1, 24, 109, 110, 125
  relativistic cosmology, 125
  special, 21
Renzini, A., 9, 18, 137
Rindler, W., 17, 18, 37, 38, 135, 136, 138
Roll, P.G., 18, 58, 137
Rowan-Robinson, M., 13–15, 18, 132, 138, 139
Rozental, I.L., 132
Ruffini, R., 116, 122, 138

Sachs, R.K., 95, 96, 138
Sakharov, A.D., 68, 70, 138
Scalar boson, 107
Scalar inflaton field, 71, 124, 125
Scale free spectrum, 79, 80, 83, 124
Schwarzschild radius, 110, 119
Sciama, D.W., 15, 17, 18, 132, 139
Shapiro, S.L., 116, 121, 137
Shapley, H., 14, 18, 138
Silk damping, 100, 104–107, 124, 126
Silk mass, 101, 108
Silk scale, 99–102, 107
Silk, J., 17, 18, 62, 65, 95, 96, 99, 100, 108, 133, 138, 139
Singularity, 52
Singularity theorems, 29
Slipher, V.M., 6, 7, 38
Songaila, A., 16, 19, 138
Speed of light, 36, 127
Spinless boson, 72
Statistical equilibrium, 62
Statistical fluctuations, 79, 80, 124
Stauffer, D., 100, 108, 139
Structure, see Large scale structure

Taylor expansion, 40
Telescopes, 5, 6
  optical, 5
Thermal radiation, 52, 53
Thomson cross-section, 100
Thomson scattering, 56
Tipler, F.J., 8, 18, 132, 139
Toussaint, D., 68–70, 138
Treiman, S.B., 70, 138
Tritium, 62, 63
Turner, M.S., 17, 18, 112, 114, 115, 121, 122, 133, 138, 139

Unified theories, 12, 112
Universe
  age of, 8–10, 17, 30, 31, 35, 52, 59, 78, 95, 103
  history of, 94
  large scale homogeneity, 14–15, 81, 83
  thermal history, 2, 49–58

Vandenberg, D.A., 9, 19, 138

Walker, D., 18, 138
Wasserman, I., 116, 121, 137
Wavelength, 35
Weak interactions, 54, 55, 64
  freeze out, 57, 64, 108
  rates, 60, 61
Weinberg, S., 8, 10, 17, 19, 29, 31, 40, 41, 47, 58, 69, 70, 90, 96, 131, 135, 138, 139
Wilczek, F., 70, 132, 138
Wilkinson, D.T., 18, 58, 137
Wilson, R.W., 18, 52, 58, 138
Wolfe, A.M., 95, 96, 138
Wolfram, S., 69, 70, 137

X-rays, 111

Yahil, A., 18, 138

Zee, A., 70, 138
Zeldovich, Y.B., 10, 19, 79, 84, 117, 121, 135, 137–139